Lecture Notes in Control and Information Sciences

Edited by M. Thoma and A. Wyner

172

H. Tolle, E. Ersü

Neurocontrol

Learning Control Systems
Inspired by Neuronal Architectures
and Human Problem Solving Strategies

Springer-Verlag Berlin Heidelberg GmbH

Authors

Prof. H. Tolle
FG Regelsystemtheorie und Robotik
TH Darmstadt
Schloßgraben 1
6100 Darmstadt
Germany

E. Ersü
ISRA Systemtechnik GmbH
Mornewegstraße 45
6100 Darmstadt
Germany

ISBN 978-3-540-55057-0   ISBN 978-3-540-46680-2 (eBook)
DOI 10.1007/978-3-540-46680-2

Typesetting: Camera ready by author
61/3020-5 4 3 2 1 0 Printed on acid-free paper.

To
Professor Dr.-Ing. Dr.-Ing. E.h. W. Oppelt
and
Professor Dr.-Ing. W. von Seelen

who introduced us to neural information processing
and supported our work in this area by many helpful discussions.

Preface

The content of this book was stimulated by a lecture, given by Professor Dr.-Ing. Dr.-Ing. E.h. W. Oppelt in connection with the 100 year anniversary of the TH Darmstadt in October 1977, which put forward the idea, that technical controllers can be considered as an artificial intermediate device between dead matter and living creatures - /1/. In some deeper discussions on the differences and deficiencies of technical controllers compared with the abilities of the human brain, Professor Oppelt introduced me to the actual status of brain and neuronal network theory and the most important literature on these subjects.

The study of the literature aroused my interest to find out how far relatively advanced brain models could be used to simplify the controller synthesis for heavily non-linear and/or complex and difficult to model technical processes, since control theory is fairly complicated, even for those processes which can be handled very easily by human beings. There were, naturally, other scientists, who had started to explore this idea much earlier. However, the technical aids for such work were very often not adequate for such work then. E. g. K. Steinbuch had to build up his model of a neural network, the learning matrix, using relays, and was thus limited to a very small network - /2/.

In 1978 I met three very positive aspects: (1) Cheap memories and parallel processors, being basic aspects of the human brain, started to become available. (2) J. S. Albus had developed in /3/ a very attractive model of some neural network, in which the basis was not the sensory aspect, as in most earlier pattern recognition neural network models, but rather the stimuli-response requirements of the cerebellum, which means adequate action generation for a certain situation as in control tasks. (3) In E. Ersü I found a researcher who was as enthusiastic about the research goal as I was, and who was, in addition, an extremely good and creative scientist, having furthermore the gift to motivate students to work in this area.

Actually, the book is a compilation of research work directed mainly by E. Ersü, refined by J. Militzer, and continued by M. Hormel and W. Mischo in recent years. Of the order of 30 students contributed to the results by their study and thesis elaborations. Although I cannot enumerate them all here, I would like to thank them for their full and innovative cooperation.

The work has been supported partly by the Deutsche Forschungsgemeinschaft (DFG) and the Stiftung Volkswagenwerk and it has also been shaped to and tested on industrial processes by cooperations with the Robert Bosch GmbH and the Bayer AG. All this aid I would like to acknowledge here.

To avoid wrong expectations and misinterpretations of the content of this book, some remarks on modelling may be helpful, especially since in the recent years neural information processing based on neural net simulations has become a very fashionable research topic. At first one has to consider the goal of the modelling. In natural sciences this goal is to achieve a deeper understanding of our world: fairly general laws hypothesized on experimental results are more and more refined over time. Very often clearly different levels of modelling depth are found to be appropriate descriptions for specific aspects of the microcosmos and macrocosmos . Examples from physics and neural information processing are given in table 1.

- Newton's laws, gas laws....
- atomic/nuclear physics

- elementary particle physics

- psychological laws of behaviour
- models on neuronal level
  (e.g. brain sections, neuronal activities)
- electrochemical membrane activities

Table 1 - Examples of modelling levels, left side from physics, right side from neural information processing

In contrast to this analytic task in the natural sciences, in which always new, not understandable aspects lead to further experiments with the aim to clarify further details, the goal of engineering is synthesis, which means using as far as possible existing knowledge to build systems for an improved control of nature for the sake of facilitating human life. As far as control engineering is concerned, experience shows that for modelling with this purpose in mind the level of Newton's laws is satisfactory in general. Very often even some rougher models - reductions of so- called state space models - suffice. Keeping this in mind, one should expect that engineering applications of neural information processing can work mainly with the results of knowledge about the psychological laws of behaviour and/or condensations of findings from the neuronal level into similar crude general laws.

Actually, this has been proven to be true in our work. So the book only makes use of abstractions from findings on the neuronal level and of ideas about the human behaviour. Details with respect to the neuronal level are given just as far as necessary to show how the concepts put forward can be motivated from results in this area. On the control loop level the basic structure was designed by E. Ersü before a careful comparison with the literature showed that it was in good accordance with human behaviour.

I would like to conclude this preface by thanking Professor P. C. Parks from the Royal Military College of Science, Shrivenham, England, for his cooperation on convergence and stability problems during his year as guest professor 1986/87 in Darmstadt, and also afterwards and for his help to me in writing this book in hopefully understandable English. This book would not have been possible without the willingness of Professor M. Thoma to include it in the lecture note series, and the kind and always responsible way, in which my secretaries, Mrs. S. Beyer, Mrs. H. Noparlick and Mrs. G. Tröger accepted the respective additional work load. They deserve my thanks as well as cand.-ing. C.-M. Marquardt and cand.-ing. H. Schomaker for their careful drawing of the many figures in this book and Dipl.-Ing. T. Sticher for arranging the final layout of the manuscript.

E. Ersü has been asked by me to act as co-author, since this book is based to a high extent on ideas developed by him and on student work supervised by him. However, I take the full responsibility for the contents of this book.

Darmstadt, April 1991

/1/   Oppelt, W.: Der Automat - das Denkmodell des Ingenieurs für menschliches Verhalten. Die Elektrotechnische Zeitschrift (etz-a) 1978

/2/   Steinbuch, K.: Learning Matrices and their Applications. IEEE EC 1963

/3/   Albus, J. S.: The Cerebellar Model Articulation Controller. Trans. ASME ser. G. 1975

Contents

I. Basic considerations

I.1. Task description

With the availability and growing cost effectiveness of the digital computer more sophisticated controller designs have become possible. This is of special relevance for controlling heavily nonlinear and/or very complicated processes. Two major lines of development have evolved:

a) Process tailored controllers: Through intensive physical modelling a very detailed process description is set up and the controller - in general a mathematical algorithm with nonlinear elements implemented on a microprocessor - takes into account the whole complicated and/or nonlinear behaviour of the process. By this means highly improved performances can be achieved for well defined systems like electric motors - see e. g. /1/ -. A further advantage is that physically correct models allow fault diagnosis and prevention through intelligent measurement monitoring - see e. g. /2/ -. However for certain plants, like biochemical processes, detailed modelling is still very problematic. In other cases generating specific mathematical models may take too much time or, may seem to be too costly. In addition specialist manpower is required which especially in smaller companies is either not available or not trained in the art of transferring process knowledge into mathematical models and into algorithmic controller design. Also specifically tailored nonlinear controllers are normally not robust with regard to differences between the process models and real plant characteristics. Such differences may arise e. g. from not precisely known parameter values, unmeasurable permanent disturbances or unpredictable ageing. Lack of robustness means that small differences between the process and its model lead to large performance degradations and - in the worst case - to instability problems.

b) Parameter adaptive control: Instead of developing a detailed analytical model of the process from physical considerations one tries to use a simple local description of the process behaviour derived experimentally and being just sufficient for control purposes. The nonlinear aspects are covered by adaptation of the model parameters to the actual process working condi-tions. The algorithms are installed on a dedicated microcomputer or some other computer with further job assignments and work in general in the following way - for more details see e. g. /3/ -: A low order linear process model structure is estimated or automatically determined - experience has shown that one can live mostly with a third order model plus a deadtime - and a set of initial parameter values is selected. These values are shifted then iteratively into an appropriate set by on-line stimulation of the process and the model with rich enough input signals and a comparision of the respective outputs. A suitable linear controller - chosen according to the design goal - is rapidly adapted to the identified process model by the respective analytical algorithm. So with each change of the working conditions - e. g. a set point change - new parameter sets are generated for the process model and the controller, leading to good closed loop behaviour under all working conditions. However the crude process model used is not an image of the real physical process and the system is always active even if the process returns to situations already encountered. This means that the same amount of computation to select the right controller has to be made available, independently of how long the adaptive control system has been installed.

Both these engineering solutions to control complicated and/or heavily nonlinear processes are not fully in accordance with the behaviour observed when human beings handle such tasks. On the one hand human beings seem to get away with rather imprecise models and control algorithms for achieving fairly robust control procedures. On the other hand the process models are sufficiently detailed that a continuous adaptation to changing situations is not necessary: for situations already encountered the appropriate controller has not to be generated anew, but is remembered, so that the attention of higer levels of the brain is not required after a while for these commonly encountered situations. Furthermore human models and control algorithms seem to include nonlinear elements if required.

Starting from this state of affairs the goal of the research to be discussed in this book is, to investigate a new additional solution for the control of heavily nonlinear and/or complicated technical processes, inspired by current ideas on human strategies to handle control problems and on the underlying hardware, namely the brain.

The interesting aspects of this approach are that it keeps the flexibility of adaptive control, that means it can be applied without major changes directly to different processes. In addition one gets a detailed process model representative for the whole set of working conditions, making recomputations for known situations superfluous. The controller derived automatically from this model is a nonlinear tailored one leading to very good performance, but it is still relatively robust through the attribute of local generalization, a concept to be introduced later on. A shifting from computational effort to memorization connected with this concept may allow furtheron the generation of a very large scale integration (VLSI) solution with additional advantages for the control of very fast processes and in savings of space necessary for controller hardware.

Finally, it has to be pointed out, that although the ideas have been derived from solutions applied by nature it is not the objective of the research work to be discussed in this book to analyze and imitate these ideas in depth. Any alternative approach to classical control theory leading to good results is of interest, whether or not it models living systems exactly. Therefore only a brief discussion on a relatively high, functional level regarding neuronal networks and human strategies in dealing with the environment will be given, and mainly mathematical analysis of the resulting structures is considered.

I.2. Information processing and memorization in neuronal networks (microintelligence)

Information processing in animals is performed in so-called neuronal networks with the human brain as its highest development. Neuronal networks are cellular arrangements, being built from a certain variety of cells (neurons) which, however, all have the same basic features - see e.g. /4/ -. Each cell has a nucleus embedded in a cellular body with a cell membrane as its skin and extensions, through which it collects incoming signals and distributes outgoing signals. The branches collecting incoming signals are called dendrites, the branching extensions through which outgoing signals are distributed are called axons (fig 1). The axon branches are connected to the dendrites and/or the cellular bodies of other neurons by so-called synapses (fig. 2).

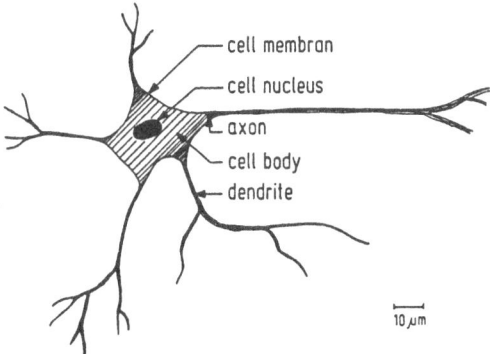

Fig. 1: Principle cell structure (from /4/)

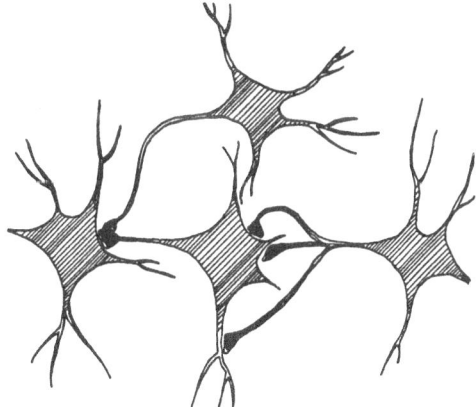

Fig. 2: Examples of synaptic axon end connections (from /4/)

The differences between different neurons express themselves mainly in the way and degree of dendrite branching. The information transport itself is fairly complicated, being performed on the axon by electrical currents triggered through potassium - sodium ion exchanges in the cell membrane; it will not be discussed here since we are only interested in the possible general functions of large groups of neuronal cells.

The human brain comprises of the order of $2,5 \bullet 10^{10}$ cells (/4/) and although it has many different tasks, which are concentrated in certain areas - like sensory input processing, effector orders generation, memorization and general information processing - all areas responsible for a specific task are sufficiently large that one can consider them as big networks themselves . Fig. 3 shows an enlarged cut through the cerebral cortex, making the structure visible by a special preparation (staining) method, due to Ramon y Cajal. The extensive interconnections between the neurons through the dendrite branching motivates the expression "neuronal networks". Actually the synaptic connections have to be divided into inhibitory and excitatory connections, since some of them try to suppress and some of them try to increase the neuronal activity. The input signals at the synapses are weighted in the synapses and added up with negative values in case of inhibition and positive values in case of excitation in the respective neuron. Only if the sum is higher than a certain threshold, the neuron "fires", that means sends out an electrical current, a "spike", along its axon. By this means the overall activity of neuronal networks is kept to a certain level and cannot go up to infinity. A detailed description of the human brain and its hardware can be found e. g. in /5/.

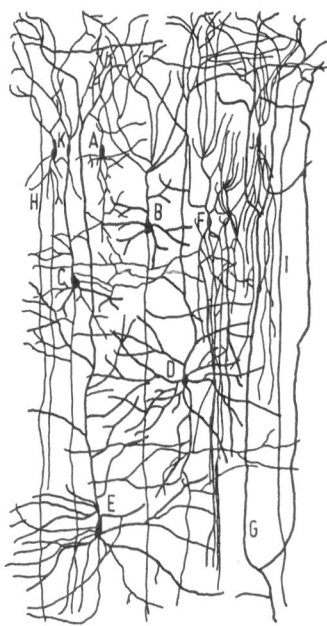

Fig. 3 (from /5/): Cell network of the cerebral cortex. A-K different kinds of cells and/or cell branching

For us the details of neuronal networks and information processing therein are not of interest, but only their ability of associative learning. By associative memorization and/or learning we mean the capacity to connect certain outputs with certain inputs and information on the circumstances (situation) under which they arrive in the memory, so that any input to the memory under the same circumstances triggers the learned (trained) output without any search being necessary. In modern artificial intelligence language one could call this an "if-then" memory rule. The hypothesis of how the associative memorization is achieved in neuronal networks is that in the training cycle synaptic connections specific to input and circumstances are reinforced, which means that the weighting in the synapses possesses some plasticity - see e.g. /6/ -. Since the overall network activity is kept at a certain level by a balance between excitation and inhibition, the reinforcement of certain connections means that the neuronal network is dominated by this path. Other possible signal paths are closed by the thresholds, and one obtains the learned output.

For inputs which are very similar to the learned input under the same or very similar circumstances one has to expect the following behaviour: Only some of the many paths, which define in the summation of their signals the output, will be different to the activated path by the learned input/situation pattern. So the output will be very similar to the output of the trained situation in this case, although the actual input/situation pattern has not been trained. We call this "local generalization", "local" since for very dissimilar input/situation patterns none of the paths learned in the trained case will be activated in general, which means that there is no shaping of the neuronal network response through already trained input/situation patterns in this case. The high number of possible paths originally formed by chance leads one to expect further, by the way, that damage to the network may diminish the accuracy of the output, but would not - if not being too big - lead to wrong outputs for the learned input/situation pattern.

The freedom of storing arbitrary input/situation - output relationships for clearly separate input/situation patterns turns neuronal networks into devices, which are able to store general mappings between say n dimensional incoming patterns (input/situation patterns) and say m- dimensional output patterns. The large number of cells and branchings involved in each case seem to be at the first sight an extravagance; however, they guarantee a high storage capacity and - through the broad information distribution - a high safety margin against information loss caused by possible destruction of parts of the network. The local generalization leading to a certain smoothing of (automatic interpolation between) stored data reduces on the other hand the necessary memory space with the big advantage of reducing the necessary training effort at the same time. The price one has to pay for this advantage is that relatively different output patterns for relatively similar incoming patterns are difficult to achieve; a very high training effort is necessary for such behaviour separation.

That the given description seems to be a good high level characterization of neuronal network behaviour can be seen by considering the human ability to learn special movements. If one takes, for example, sports, a lot of very different movements for performing activities optimally can be learned, if the movements are very dissimilar. However, for relatively similar tasks, e.g. playing tennis and table-tennis, one finds normally, that one can see easily, which discipline a player has learned first even if the player is very good in both areas. That means, that the pattern learned first still dominates and cannot be modified into the path for the second activity which probably would have been reached as the optimal one, if only the second activity had been learned. Finally, one has learned from brainlesions that partial destructions of areas responsible for certain brain activities indeed diminishes the accuracy of the outputs, but does not result in completely wrong and/or zero outputs.

One can say that the general input/situation associative storage capability of neuronal networks, whith their ability to generate sensible answers if some similar experience is available but without the feigning of sensible answers if no similar experience is available, is an intelligent way of information processing and memorization. We shall call therefore these features "microintelligence".

I.3. Human problem solving strategies with respect to actions (macrointelligence)

To achieve the human ability to explore, understand, control and develop further the world around him the intelligent memorization in neuronal networks surely does not suffice. It has to be imbedded into some structure and general programs, which allow a goaloriented utilization of the mircrointelligence. We shall call this superior organization in the brain "macrointelligence".

The high flexibility of human activity and thinking leads to the expectation that the macrointelligence has no direct hardware background, which means it cannot be derived from brain structures directly, but has to be deduced through observation and analysis of human behaviour. So the adequate background seems to be formed by the relevant theories of psychology.

A very interesting discussion of our topic can be found in the book "Psychology of Intelligence" by J. Piaget (/7/). Piaget considers, what we call macrointelligence, as an evolving system of intelligence levels, which grow out of each other in a uniform way, having instinctive behaviour and acquired habits as the lowest and non-intelligent starting level and abstract thinking as the final, highest level. This stepwise evolution is oriented to, and backed up experimentally by, an analysis of development of intelligence in human beings, beginning with the first expressions of intelligence in the infant and following the changes in quality in the years in which the individual grows up and builds up his full mental powers.

Without going into details it should be remarked, that Piaget distinguishes two main standards of intelligence: The so-called "senso-motoric intelligence" enables the individual to combine perceptions and actions into a goal orientated behaviour. "Thinking", however, allows abstract considerations and conclusions, not necessarily orientated to an actual request for some action. Since, due to

Piaget, intelligence is formed in a step by step enlargement of its abilities, the boundaries between senso-motoric intelligence and simple forms of thinking are not sharp and therefore a number of similarities and interconnections between both standards exist.

However, in connection with modelling macrointelligence it seems to be adaquate up to now, to use different tools for modelling abstract thinking and senso-motoric intelligence. For abstract thinking the methods developed under the heading of artificial intelligence, like information storage in a tree-like structure and search of this structure by heuristic driven, context-restrained procedures (see e.g. /8/ - /11/) form a very interesting approach which is already very helpful in the area of expert systems. Our research does not deal with this aspect, however, but concentrates on the senso-motoric intelligence.

By senso-motoric intelligence we mean goal-orientated behaviour in connection with directly necessary actions without raising the action generation to the level of reasoning with symbols and/or linguistic characterizations.

It should be remarked in this connection, that the formation of an abstract characterization of some perceived object, allowing the symbolic handling of an object, is an open question and that this missing link between "senso-motoric intelligence" and "thinking" forms one of the main difficulties for further understanding and modelling human intelligence. A respective scheme is especially of importance for an eventual combination of AI-methods and the learning control discussed in this book, if the AI-methods are understood as methods of a higher hierarchical level, which takes advantage of learned schemata to handle certain situations in an optimal way.

Concentrating on the "senso-motoric intelligence", their description can start from the following statement. (/7/, p. 108). The complete intelligent action needs three elements:

- the question, which directs possible search actions

- the hypothesis, which anticipates eventual solutions

- the control, which selects the solution to be chosen.

E. Ersü, whose control loop models form the basis of this book (and were developed without considering the ideas of Piaget) translated 1983 these requirements into the following elements for a control loop imitating the senso-motoric intelligence (/12/).

- the "question" is equivalent to a performance criterion, which allows in a mathematical environment to assess the advantages and/or disadvantages of possible actions.

- the "hypothesis" is equivalent to some predictive model of process answers (predictive model = a model giving the process status at the next instant out of the present status and possible inputs), allowing to test possible process inputs for the current situation with respect to the performance criterion, without the necessity to apply these inputs in reality

- the "control" is equivalent to a device creating and performing the control action to be applied in reality.

Ersü remarks in this context, that one further requirement is necessary, namely

- the ability to build up the predictive model mentioned above.

This translation of the psychological characterisation of intelligent behaviour into engineering thinking leads to the overall structure of fig. 4.

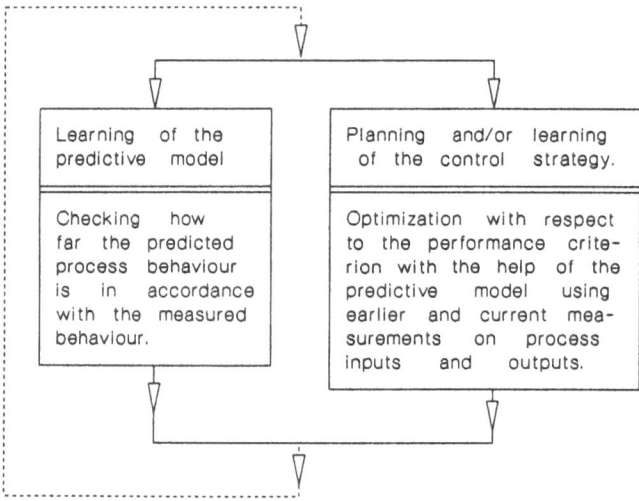

Fig. 4: General scheme of a learning control loop imitating senso-motoric intelligence (/12/)

To be more specific one has to distinguish between two different levels of senso-motoric intelligence, depending on the level of sophistication used for the predictive model. In the case of the less sophisticated one, the "empirical intelligence" from /7/, the results of senso-motoric tests are only judged and stored away regarding the quality of being advantageous or not. In the more sophisticated case, the "systematic intelligence" of /7/, the input-output relationships are used to build up a model of the environment. This has the advantage, that for a certain input one gets not only the answer, whether the result would be good or bad, but also what would be the status of the process then, so that one could apply a further input and get a further forecast and so on. Thus, such an explicit process model allows one to a certain extent long range planning of control strategies.

However, for simple tasks or whenever the building up of an explicit model requires too much effort, a certain optimization of process handling can also be achieved, by always using the best input for just the next step, which requires only storage of the performance criterion value attributed to such inputs under the current situation. This "empirical intelligence" approach with its implicit process behaviour modelling safes storage place and learning effort at the price of slower and less accurate learning of optimal behaviour. Both kinds of intelligence have been studied and will be discussed later on.

One further point to be mentioned on senso-motoric intelligence are the so-called implications from /7/: Those are certain preferred search actions in the case that not enough has been learned about the process for the performance criterion to indicate the best actions. Those "implications", which are generated through a "mysterious mechanismus" (Piaget), are taking into account general relations known about the process and are normally used to extend the knowledge about the process. If this does not lead to a success, a random search is used to improve the process model. Actually, it is important to learn at the same time good control strategies and new process details as quickly as possible and we shall meet later on in this book strategies to do this under the heading "active learning", which is a mathematical scheme to generate the "mysterious" implications.

Finally it has to be stressed, that speculations on the software structure of the brain are much harder to support experimentally than the ideas on possible functions of hardware connections, like the basic assumptions regarding the performances of neuronal networks. Also one should always remember, that with the senso-motoric intelligence only a small segment of the overall intelligence is taken into account. So not everybody will be satisfied with the correctness and the degree of modelling used in this book regarding the "macrointelligence" aspect.

However, since technical control loops have as their basic requirement the generation of favourable actions with respect to a certain goal on the basis of information generated by sensors (measurements), the restriction in modelling the senso-motoric intelligence seems adequate as a first step for imitation of human intelligent behaviour. Again the biological correctness is only a desirable feature to have, the main point for the engineer being how far by such an approach, new and improved possibilities for control loop design are generated.

I.4. The learning control loop due to E.Ersü

In 1980 E. Ersü proposed in an internal report the combination of locally generalizing associative memories and what we have called senso-motoric intelligence with an explicit world model (the systematic intelligence of Piaget) as a new control concept for highly nonlinear technical processes. The mathematical details will be discussed later on in this book. Fig. 5 gives, however, the general structure as presented at the German Biocybernetic Congress in 1983 - /13/ - with some adaptation of the block characterization by using engineering instead of biological expressions.

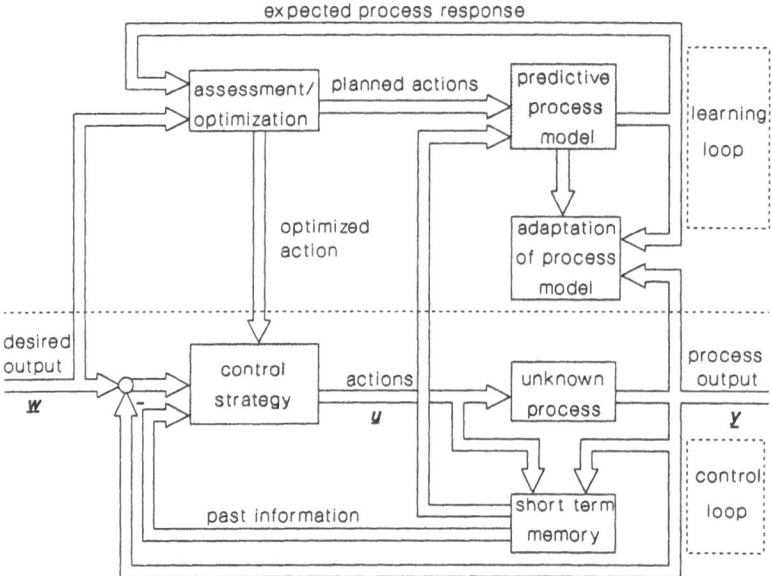

Fig. 5: Learning control loop with associative memory systems (shaded blocks)

Comparing this lay-out with fig. 4, one can see, that all elements of the "systematic" macrointelligence scheme can be found: a predictive process model, a control strategy block, an optimization procedure and some adaptation element for updating the process model. The required learning ability regarding the predictive process model and the control strategy are taken into account by using microintelligent elements - locally generalizing associative memories - at the respective places in the macrointelligence scheme. The current situation of the unknown - or better not very well known and/or mathematically modelled - process is described by its inputs, its outputs and possibly some history of the inputs and outputs provided by a simple hold element, the "short term memory" in fig. 5. Actually, the control concept of fig. 5 is implemented with the help of a digital computer, leading to a sampled data system, in which, - kT being the actual time, - the history means inputs and outputs at times $(k-1)T$, $(k-2)T$...(T = sampling period), which can be made available in the most simple manner by just remembering those values.

In some more detail, the scheme of fig. 5 functions as follows:

The control actions are commanded by the associative memory block "control strategy", which is taught advantageous control actions for certain situations (process situations plus deviations between actual and desired process outputs) by the learning level. If the control strategy block has learned good control actions for enough different situations, the control loop can run in principle independently from the learning level, creating fairly correct control actions for situations not directly met by automatic interpolation through local generalization. That means, that after a learning phase - in which if desired one can work in a slower mode as in the real application or even off-line with some process model instead of the real process - there is no longer any necessity of computations, but there is a direct generation of advantageous actions by memorization of the respective control strategy for the actual (measured) situation. This imitates human abilities e.g. in sports, where movements are learned in a slowed down mode and are repeated later on much more quickly, as would be possible if one had to reason about the appropriate movement.

The learning level is responsible for building up the information on advantageous control actions. It does this with the help of the associative memory block "predictive process model", in which the plant output at time $(k+1)T$ is stored for certain situations at time kT. By testing different possible inputs for the current situation through applying them to the predictive process model, one can decide with the help of some performance criterion (block assessment/optimization) which one is the best one with respect to the desired performance and can transfer this plant input to the associative memory control strategy, where it is stored together with the respective situation. Using a locally generalizing associative memory for the predictive plant model leads as with the control strategy again to an automatic interpolation between certain discrete information points, generated through the specific situations met earlier.

The whole system is a self-organizing one in the sense, that at first there exists neither a predictive plant model nor any control strategy, which means we have empty or randomly filled associative memories. So at first, some completely incorrect actions will be produced together with some completely wrong predicted plant outputs by the predictive plant model. However, the adaptation block (fig. 5) will correct the predictive plant model in a step by step procedure, so that after some time a fairly good prediction of plant behaviour is possible and also by then a selection of good control actions to achieve the desired plant output. As with human beings, satisfactory learning requires frequent repetition of the same effort,so the behaviour of the system is improved by being restarted from the same initial conditions again and again. Important for technical control problems is the ability, to stabilize the control loop already in the first trial, however, with relatively bad performance in general. Fig. 6 shows the general layout of a pH-neutralization stirred tank reactor, which will be used to demonstrate this behaviour. Again, the discussion of physical and mathematical details is delayed to later parts of this book. The main nonlinearity, which makes this process difficult to control, is the so-called "titration-curve", connecting the pH-value with the ionic concentration of the acid to be neutralized. This is, as shown in fig. 7, a logarithmic function.

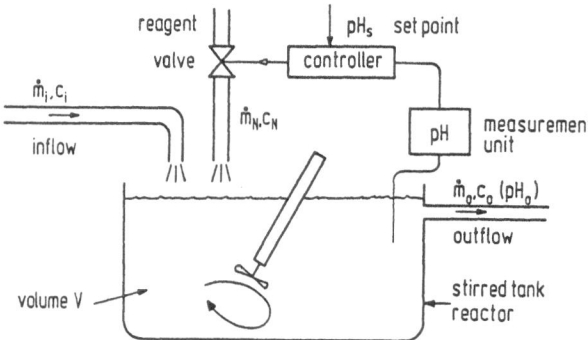

Fig. 6: General layout of pH-neutralization. c_i, c_o, c_N: inflow, outflow, reagent ionic concentration, $\dot{m}_i$, $\dot{m}_o$, $\dot{m}_N$: inflow, outflow, reagent mass throughput

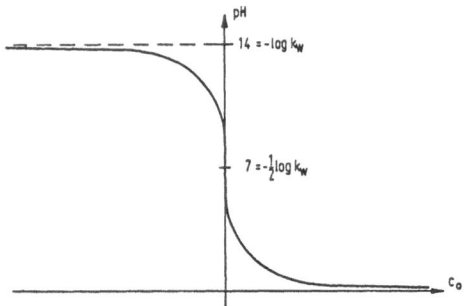

Fig. 7: Titration-curve; c_o ionic concentration of the acid to be neutralized, k_w = dissociation constant of water, pH = $-\log [H^+]$, H^+ = hydrogen ions

The behaviour described above can be read out of fig. 8 and fig. 9.: The first start from some initial conditions with the demand to reach a pH-value of "9" is marked "I". Fig. 8 shows the plant output, fig. 9 the plant input. One sees, that in the beginning the plant input is switching between zero and some growing value until through this information generating effect - being determined by an active learning strategy not discussed here - enough knowledge is acquired to raise the control level without intermediate switching to zero. After some period, the requested pH-level is reached furtheron. However the overall procedure cannot be judged to be very satisfactory. This is achieved stepwise by repeating the same control problem, that means by starting from the same initial conditions with the demand to reach the same pH-level. Due to the fact that the predictive plant model has now already information on plant behaviour, one gets a better and better transfer from the initial conditions to the value pH = 9.

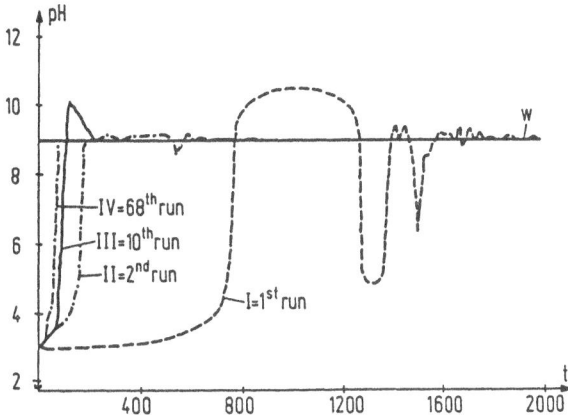

Fig. 8: pH values reached by a different number of runs in sequence from the same initial conditions with no plant and control strategy information in the respective associative memories in the beginning

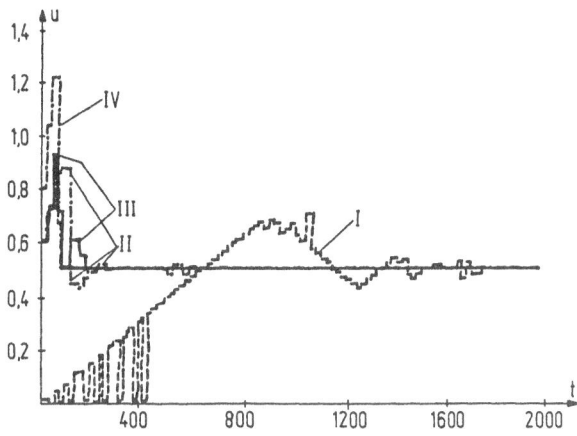

Fig. 9: Control actions u(t) generated by the learning control loop

I.5. Comparison of adaptive and learning control

Based on the given description of the learning control loop considered by us we are now able to deepen the general discussion from "I.1 Task description" on the differences and similarities between this scheme and more conventional adaptive control which is the most similar existing control concept.

Fig. 10 is a high level block diagram for adaptive control as can be found e. g. in /3/. It is drawn up for a single input/single output process and adaptation as well as memorization of earlier signal values to be used in the process model are included in the block identification and so on. As has been stated already in I.1, in general one uses a linear process model together with a suitable linear controller, the parameters of both being determined through the identification and the controller calculation based on this identification. Taking into account, that this is also a computer implemented digital control system and neglecting the deadtime element very often necessary to model the process efficiently, the applied difference equations read for the process model (the sampling time T being suppressed in kT, (k-1)T...):

(1)
$$\tilde{y}(k) + a_1\,\tilde{y}(k\text{-}1) + \dots + a_n\,\tilde{y}(k\text{-}n)$$
$$= b_1\,\tilde{u}(k\text{-}1) + \dots + b_m\,\tilde{u}(k\text{-}m)$$

and for the controller:

(2)
$$\tilde{u}(k) + \alpha_1\,\tilde{u}(k\text{-}1) + \dots + \alpha_\nu\,\tilde{u}(k\text{-}\nu)$$

$$= \beta_0[w\text{-}\tilde{y}]\big|_k + \beta_1[w\text{-}\tilde{y}]\big|_{k\text{-}1} + \dots + \beta_\mu[w\text{-}\tilde{y}]\big|_{k\text{-}\mu}$$

$\tilde{y}$ and $\tilde{u}$ are the deviations of the absolute signals y, u from the steady state or d. c. (direct current) values y_{00}, u_{00}.

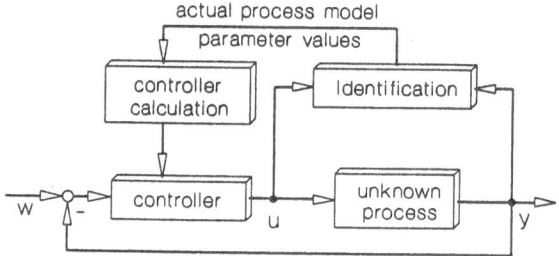

Fig. 10: Basic scheme of parameter adaptive control (see e. g. /3/)

The dynamic process model is - in contrast to the dynamic controller - predictive in the sense that from knowing the $\tilde{y}$ and $\tilde{u}$ values at times (k-1)T, (k-2)T... one can calculate a plant outpout estimate $\tilde{y}$ at time kT. However, the process model is not used in this way. Representing a hyperplane through the origin in the space spanned by $\tilde{y}(k)$, $\tilde{y}(k\text{-}1)...\tilde{u}(k\text{-}m)$, its parameters a_1, a_2.. b_m are calculated as to give for a certain set of measured $\tilde{y}$-, $\tilde{u}$-values the minimal quadratic deviations from the hyperplane. This is in effect an estimated tangent plane for y_{00}, u_{00} to the in general non-linear process in y(k), y(k-1)..u(k-m)-coordinates. The translation between the y-, u- and the $\tilde{y}$-, $\tilde{u}$-coordinate systems is defined by the steady state values. y_{00} is in the closed loop in general represented by w and u_{oo} can be calculated by using y_{oo}=w through including an additional constant into equation (1) and estimating this constant together with the parameters $a_1..b_m$. For further details see /3/.

The linear process model determined is used to calculate an adequate linear controller of the form of equation (2) by linear control theory. As long as no new identification is made (new set of parameters a_1-b_m calculated from further measurements) this controller - and by this implicitly this process model - is applied globally, that means independently of the question of how far the actual situation is away from that described by the measurements used for getting the linear process approximation. We can call this "global generalization".

In contrast the "local generalization" of microintelligence works as follows: For each measured value the result is expanded e. g. into a cone with its peak value identical to the measured value and its slope being a parameter characteristic to the defined associative memory which means a free parameter of the memory selected by the user from experience and/or general considerations. The output of the associative memory for a certain input is the adding up of the respective values of the different cones stretching over the point considered. This leads to an automatic interpolation between different measurement points. However some adulteration occurs - minimized by the distributed storage principle to be explained in chapter II - in the case of training of nearby points.

In fig. 11 the results of "global generalisation" and "local generalisation" are set against each other for some measurements, without taking into account the unavoidable measurement noise. Further-on a reduction to a two-dimensional case has been used, although the most simple realistic process model would consist out of y(k), y(k-1) and u(k-1), leading to a three-dimensional space.

The detailed discussion is given in the text under the figure. The main message is, that local generalization gradually approximates any non-linear characteristic with automatic interpolation if the points are close enough, but with the information "no knowledge about the plant" if the tested situation is not near enough to the measurement points. This latter quality is achieved by an earmarking of the respective address in the memory, when some information is stored there.

The global generalization of adaptive control can model the plant in the limits of the prescribed structure only, which means in the case of fig. 11 by straight lines only. It must switch therefore between different models even if enough measurements are available to characterize the full non-linear behaviour. Each current model pretends furthermore to be able to give an answer for all situations, including situations not met already. This may be a helpful assumption in the beginning when only few measurements are available. However, it may give also completely incorrect answers leading into a catastrophe.

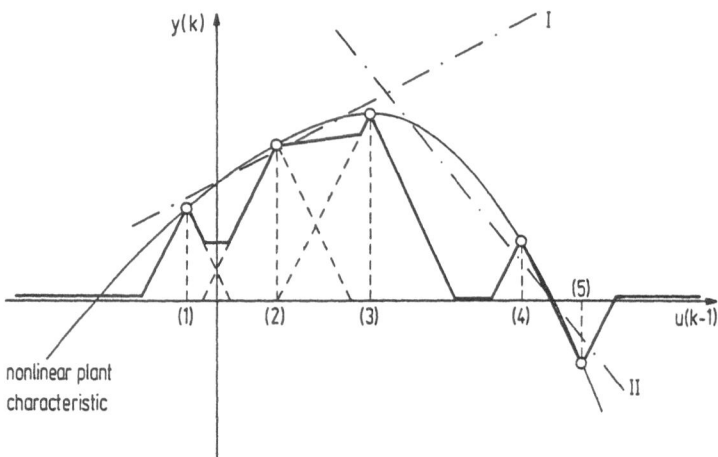

Fig. 11: Rough sketch of differences between adaptive global generalization and learning with local generalization: Measurements (1) to (5). I = first global generalization from measurements (1) to (3). There exists a prediction P for measurement (4), which is completely wrong. Only the new global generalization from (3) to (5) gives a satisfactory behaviour for y (k) in this area. However, this model is now wrong if one comes back to (1). The local generalization starts from some value in the memory, here y(k) = o for all u(k-1). (1) to (3) gives with the traced line T, summarized from the broken lines, some approximation of the plant behaviour getting better if more points are available (not possible to be shown here, since the summation procedure is simplified). For (4) it would indicate "not trained" as long as there is no measurement available later on the traced line would be extended as shown. Coming back to the situation (1) the exact value of the earlier measurement would be estimated as the plant output.

As has been mentioned already, in the learning control loop the level of the predictive process model is superfluous after a while (as long as the plant is time-invariant and no major unmeasurable permanent disturbances create a new plant behaviour, which has to be learned.) With adaptive control we have seen, that in general each set point change leads to a new process model and by this to a fresh calculation of controller parameters. So the next difference between adaptive and learning control is, that through the creation of a non-linear controller, which is advantageous for all situations met by the non-linear plant, the permanent computation required by adaptive control is replaced by a suitable memorization of good outputs for the current situation. This may lead to a much faster generation of the outputs than is possible with the computational load of adaptive control.

But one has to pay a certain price for this advantage. Since in adaptive control the use of history (values at (k-1)T (k-2)T...) is part of the calculation, history leads to additional inputs in the learning control concept (see fig. 5). That this remains a handable problem is however indicated by experience in adaptive control: For closed loop adaptive control one can achieve good performance in most cases already with a third order plus dead time process model and consequently with some low order controller. Since for a non-linear process model and a non-linear controller the necessary history to be included cannot be expected to be greater, but may be even smaller than in the case with linear structures, one could conjecture, that one can live in general with very little history i. e. with a small number of inputs. And this is just what we have found, when working with learning control loops.

Finally a necessary condition for identification in adaptive control is the generation of a rich enough input signal to the process (a persistently exciting input signal, /3/). Otherwise one does not get enough knowledge about the process to model it correctly. With the learning control loop one gets the process model also by using some process stimulations, however, only in a step by step procedure through the local generalization. Evaluation of possible actions using this model are valid only as far as the respective situation is learned already. So normally the next process input in the learning phase would always stay in the field of learned situations, not generating more additional process knowledge than the local generalization can give, if a point at the border of the already known process model is excited. The knowledge extension by local generalization works, but is much too slow for effective process model building. Active learning now selects, in a suitable direction in the process model space, a point some distance outside the border of known process behaviour to create new process knowledge, if otherwise a point at this border would have been arrived at. The direction is e. g. the gradient direction for the performance criterion at the border point, the amount of exploration into unknown territory being a parameter to be selected by the user, again based either on experience or on general reasoning about the process behaviour.

So active learning is substituting in a certain sense for the persistently exciting input signal of adaptive control. On the other hand it is an imitation of human curiosity, which stresses, that technical solutions require the same attributes as have been found necessary in nature for living creatures to handle/control their environment.

I.6. Short review of related work

To my knowledge, there is no exactly similar approach to our line of research to be found in the literature. However, a relatively broad spectrum of work exists with similar goals or subgoals. It is neither possible nor intended to give a complete review of these ideas here. Instead I shall discuss selected publications to give some feeling for what further possibilities exist on the different levels of our control loop design and/or what are the roots of our work.

In the first part of this discussion I should like to draw the attention to the point that the replacement of calculation by memorization to save calculation time has been considered seriously for several years already.

The second part deals with the problem of intelligent information storage and retrieval, the "microintelligence", by presenting the major impulses from the history of neuronal network modelling, and indicating in which directions these ideas have been proven to be fruitful.

Finally in the third part I shall put forward some strategies which are alternatives to the macro-intelligence concepts used in this book in order to give a broader picture of possible ideas for learning control and/or imitation of human process handling for improved plant automation.

I.6.1. Calculation versus memorization

Tabulation and interpolation between tabulated values is an often used tool, if direct calculation of the required functional solution is very complicated and/or time consuming. Logarithm tables are a well-known example from the pre-computer age. However, even with these modern devices to hand, on-line applications require sometimes the use of the tabular principle, especially, if trigonometric functions like sines or tangents have to be evaluated.

An actual example where calculation time was until quite recently undesirably large in spite of modern microcomputer hardware, is fast robot movement under sensoric supervision. For fast robot movement one has to take into account the dynamic effects. The respective calculation time, however, is responsible for the delay time between sensor measurements and possible reaction to these measurements. One would thus have for a calculation time of 50 msec and a robot movement velocity of 3m/sec a traversed distance of 15 cm, before reaction of the robot to a sensor signal, e. g. a distance sensor, is possible. Therefore the question of reducing calculation time by using tables and interpolation is especially interesting in this field.

J. Albus proposed already 1975 (/14/) to apply training and memorization - as seemingly used in motion control by the brain - to the control of robots (manipulators). However such a direct approach leads to unmanageably large memories since one needs M^{3N} storage places for it. In this formula N is the number of degrees of freedom of the manipulator and M is the number of values to be stored for interpolation between them for each degree of freedom (taken to be equal for all N degrees of freedom). One sees immediately, that even for only 3 degrees of freedom and M = 10 - normally much too low to achieve a satisfactory accuracy - one would need 10^9 memory places. Therefore B.K.P. Horn and M.H. Raibert performed a careful analysis of the governing equations and proposed a certain mixture of tabulation and calculation, which leads to an interesting reduction of necessary memory space for tabulation (/15/). A general discussion of the trade-off between calculation and table look-up in the robot case can be found in /16/. Using the ideas of automatic interpolation and learning some promising results have been reported there also (and in some more details in /17/) for a three degree of freedom manipulator using one of the investigated calculation-table look-up mixtures. That no real application seems to have taken place up to now, may be due to the following reasons:

The necessary memory space is for a six degree of freedom manipulator with a certain range of loads as a possible seventh degree of freedom very large even using the mentioned calculation-table look-up mixtures. Furthermore the inclusion of recursive schemes into the calculation of the dynamic equations - see e.g. /19/ - have reduced the necessary amount of on-line calculation so much that with modern signal-processors the calculation time can be brought down below 5 msec (see e.g. /19/, /20/). This seems to be small enough for most of the practical applications of sensory feedback. Finally, learning and storage of learned patterns instead of calculation of the respective behaviour is most appropriate for processes, which are partly unknown, so that their behaviour cannot be described exactly by differential equations etc. However the behaviour of manipulators can be described fairly well in this way, so that the learning/memorization ability of the table look-up approach is not necessary in this case.

However, in other problems of remote object manipulations, like the optimal grasping of complicated objects with a hand-like gripper, memorization of learned behaviour may be preferable to calculations based on physical constraints and/or laws. Also, if not learning for all eventual actuation possibilities is required but only for certain robot trajectories, a memorization approch may be useful and has been taken into account by different authors successfully - see e. g. /21/, /22/ -.

I.6.2. Neural network models and related theories

With respect to neural network modelling and especially neural learning, one can find a worthwile review in /23/, additional information may be found in /24/. The most influential early work in this area is that of D.O. Hebb and F. Rosenblatt.

D.O. Hebb suggested 1949 in /25/, that neural networks are able to learn through a strengthening of synaptic connections in the case of frequent similar network activations. As a result functional organisations are generated, which when stimulated, lead to certain elementary percepts. Since D.O. Hebb stated this more or less as a concept only, F. Rosenblatt tried to develop an explicit mathematical model for the ability of neuronal networks to learn responses to stimuli. This "perceptron", described first in /26/, was not intended as a detailed model of an actual nervous system but a simplification, which might or might not correspond to parts of neuronal networks and which should be helpful in the study of relationship laws between the organization of a nerve net, the organization of its environment and the "psychological" performances of which the network is capable (Rosenblatt in /27/).

The perceptron is essentially a function discriminator which may be explained most easily in connection with pattern recognition. Assume we have a white plane with a grid of lines which divide the plane into small squares usually called pixels. A certain black figure on the plane, say a character of a given size, would lead to the result, that some of the pixels would be black, others would remain white (comp. fig. 12).

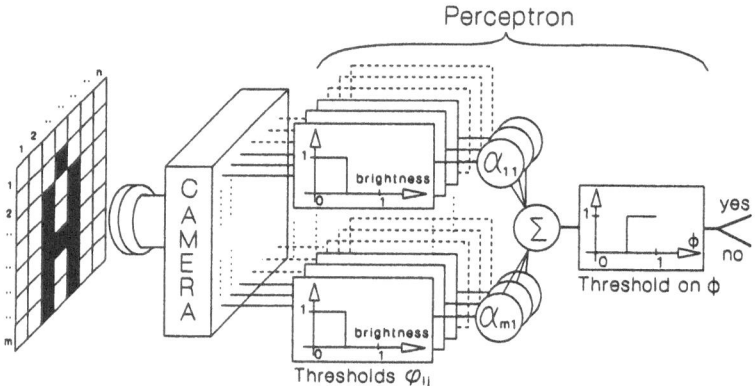

<u>Fig. 12:</u> Pattern recognition through division of picture into pixels and adding up of weighted block elements

A camera looking at the plane and classifying the pixels into white and black ones by individually assigned thresholds φ_{ij} would be able to deliver with the help of some simple computation the function

$$(3) \qquad \Phi = \alpha_{11}\varphi_{11} + \alpha_{12}\varphi_{12} + ... + \alpha_{1n}\varphi_{1n}$$

$$+ \; \alpha_{21}\varphi_{21} + \alpha_{22}\varphi_{22} + ... + \alpha_{2n}\varphi_{2n}$$

$$\cdot$$
$$\cdot$$

$$+ \; \alpha_{m1}\varphi_{m1} + \alpha_{m2}\varphi_{m2} + ... + \alpha_{mn}\varphi_{mn}$$

$$\varphi_{ij} \; \text{being} \begin{cases} 1 \; \text{for black} \\ 0 \; \text{for white} \end{cases}$$

α_{ij} being weights characterizing the pattern.

and through a further threshold on Φ a "yes" or "no" answer regarding the question, if the right points are black for a certain pattern. Naturally some training is needed beforehand, but with an adjustment of the α_{ij}-weights, one can actually recognize not only one but a large number of different patterns and split them into the classes "yes" or "no". Some defects in the learned patterns and a noisy background can be tolerated through the thresholding procedures. A sharper subdivision of patterns as by "yes" and "no" can be reached by multiple level thresholds and/or a hierarchy of perceptrons. A critical review of what can and cannot be expected from the perceptron principle can be found in /28/. There, as well as in /27/ one can pick up also a constructive proof on training convergence, which means that starting from an arbitrary set of α_{ij}-values with a certain procedure an α_{ij}-distribution can be reached in a finite number of steps, which gives a satisfactory pattern discrimination.

The initiatives of Hebb and Rosenblatt have led to further considerations in three different directions.

First, the perceptron as a pattern classifier has been improved and more directly oriented on brain structure and brain performances. A very interesting example in this respect is the model of G. Willwacher (/29/), which allows - depending on the controllable overall activity level of the system - to perform either parallel associations - complete recall of stored information when only a part of the pattern is offered at the systems input - or serial associations (meditation) - recall of a temporal sequence of patterns by input of any pattern of the sequence - two features, which are supposed to be basic processes during concept-formation and thinking in the human brain. A self-learning version of such a network was designed in /30/. There are some doubts, however, how far perceptron-like devices are useful, at least in visual information treatment, since observed visual pattern invariance principles can be fulfilled only partly by them or at a very high computational expense. Therefore some alternative ideas for this problem of information processing in the brain have been put forward, recently (see e. g. D. Marr in /31/ and H. Glünder in /32/, furtheron /30/ and especially therein the review of W. J. Freeman on the actual situation and trends in brain theory /34/).

A second line stemming from the early ideas on neuronal network behaviour is the general theory of intelligent memorization and information retrieval. An excellent treatment of this subject from an engineering point of view by T. Kohonen, can be found in /35/.

The third branch is our area of interest, which is the extension of the neuronal network modelling to the generation of commands for the motor system on the basis of sensoric information to achieve some high level goal. Two major theories for the functioning of the cerebellar cortex, where these tasks of the brain are performed, have been developed. The theory of Marr gives a general setting in which the cell(neuronal net)structure of the cerebellar cortex is analyzed and a hypothesis on the respective roles of the different elements of this structure is proposed (D. Marr /36/). No details are given for a simulation of this concept, but Marr works with a probabilistic approach, which is further described in an extension of these considerations to other areas of the brain (see /37/). The model of Albus (/38/), the second theory to be mentioned, ist fairly similar to the ideas of Marr with regard to the functioning of the cerebellar cortex. But he gives in /38/ some explicit ideas on sensory information storage including a wholly deterministic approach to overlapping information storage (local generalization), which is put into a mathematical framework in /14/, /39/. We shall discuss the details of this concept and its neuronal background in the chapter "microintelligence".

Which one of the models from Marr or Albus is of greater neurobiological relevance is difficult to say. Looking into the Science Citation Index Listings of /40/, one should assume, that the model of Marr is more widely accepted, yet M. Ito et al. state in /41/, that according to their experimental findings some of the neurobiological assumptions from Albus are more likely to occur, than the respective assumptions of Marr. For use in technical applications, there seems to exist no differences in principle between both models. However the Albus model is much more practical due to its deterministic nature. For this reason, we adopted the Albus concept as our starting point.

I.6.3. Concepts for imitation of human process control abilities

As has been mentioned already in I.3., one may distinguish at the macrointelligence level between conscious (abstract) considerations and subconscious (learned) perception-action-combinations. Accordingly different ideas for the imitation of human process control abilities have been generated.

With respect to the imitation of conscious considerations two major approaches have to be mentioned:

In the first approach only the general behaviour of humans in process control has been modelled. Some interesting ideas are discussed in /42/, where a nonlinear controller is devised, which imitates human control reactions due to the observed difference between plant output and set point and furthermore its trend. A comparison with a conventional PID controller shows advantages in the test cases considered. A much more sophisticated and very influental concept is, however, "fuzzy control" as put forward and continuously extended by L.A. Zadeh - see e.g. /43/, /44/ and /45/. By the term "fuzzy" the observation is taken into account, that human beings work to a large extent not with exact values but with relatively general statements, e.g. in the case of temperature characterization and/or adjustment, statements such as: very high, high, medium, low, etc. For processes not studied in detail or which are very complicated, this seems to be an adequate status and control characterization also, since sharper distinctions are helpful only if one can make use of them through knowledge about the process. A further advantage of the fuzzy approach is that knowledge transfer between human beings - e.g. for training of new process operators - is based on such general statements. This means that the heuristic process control rules used by the operators can be collected fairly easily, if this level of information is the only requirement for automation. R.E. King reports in /46/ on the successful operation of a rotary kiln in a cement plant by such an implementation of fuzzy control, T. Yamazaki and E.H. Mamdani in /47/ on experiences with a self-improving version of such a controller.

The second approach to be mentioned is the attempt to make use of human process handling knowledge to the greatest extent possible. The main tool in this respect are the so-called "expert systems". In these systems "if-then"-rules for process handling are set up on the basis of knowledge extracted from scientists, engineers and/or operators by questioning them about their ideas of how best to run the process. There is a similarity with the fuzzy control development as discussed in /46/, however here the control system is in general more sophisticated and its aim is more to improve overall process control than to replace operators. A discussion of experiences with the application of this idea to control problems can be found e. g. in /48/. The main problem is how to extract as much knowledge as possible about the complicated and/or only fairly well known process from the "experts". In /49/ one can find some ideas, how to improve this situation by a cyclic procedure.

With respect to perception-action combinations, the senso-motoric intelligence, one has at first to remark that concepts developed from ideas on human abilities are on this level not very far from concepts coming from pure system engineering considerations of how to handle processes which are partly unknown or difficult to model. As a unifying title, G. N. Saridis used therefore in /50/ the expression "self-organizing control". Under this heading he discusses as well parameter-adaptive control as performance-adaptive stochastic automation.

A short description of parameter-adaptive control has been given already in I.1. The main feature is, that these self-organizing control systems are based on a given structure for the plant model and the controller, and that only the free parameters of this structure are changed. Here a desired input-output behaviour of the closed loop is achieved in spite of incomplete knowledge about the process and its interactions with its environment (disturbances). In performance-adaptive self-organizing control an appropriate controller structure and eventually process model structure (if a process model is part of the closed loop system) is built up out of the observation of input-output behaviour. Such systems are "learning systems" in just the sense in which we are using this expression. However, Saridis points out that learning systems in general comprise more than the control task, as can be concluded from the following definition, worked out by a subcommittee of IEEE (/51/):

"A system is called "learning" if the information pertaining to the unknown features of a process or its environment is learned and the obtained experience is used for future estimation, classification, decision or control such, that the performance of the system will be improved."

So the expression "learning" is to be considered as a characterization for systems which try to imitate human behaviour, "learning control" being a subelement only. Self-organizing capabilities of adaptive controls can be achieved by mathematical modelling of brain hard- and software, an example being the topic of this book, as well as using the engineering ideas of controller adaptation on the basis of inputs and differences between observed and desired outputs. Thus we have here in the above mentioned field a meeting between concepts based on behaviour of higher creatures and concepts based on engineering considerations.

The most important work on self-organizing performance adaptive control has been performed using stochastic descriptions. A very detailed review can be found in the book by G. N. Saridis /50/ already mentioned and a more recent review is an article by K. S. Narendra (/52/). In line with the cited definition of learning systems, such stochastic learning automata are not applied to control problems only, but also to general decision problems. The telephone routing problem studied by K. S. Narendra and his co-workers (see e. g. /53/) can be mentioned here as an example.

There are a number of reasons that hierarchical structures exist in the hard- and software of higher creatures. R. P. Hämäläinen reports for instance in /54/, that a major step forward in modelling of breathing control was possible only by assuming a hierarchical structure. J. S. Albus discusses in /55/, that from experiments with animals and observations of injured humans one can conclude that different levels of complexity in the sensor-motoric behaviour are blocked depending on the area where the brain is cut. This can be explained only by the assumption of a hierarchical control structure (see fig. 13).

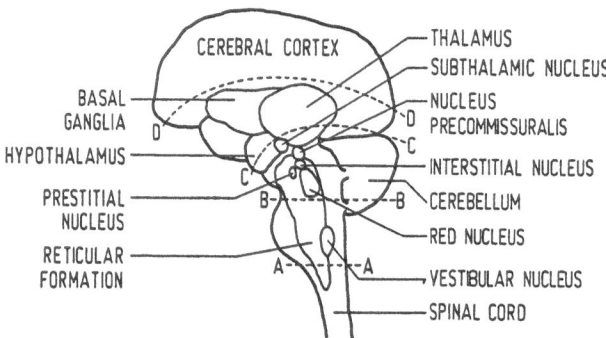

Fig. 13: The hierarchy of motor control that exists in the extrapyramidal motor system. Basic reflexes remain even if the brain stem is cut at A-A. Coordination of these reflexes for standing is possible if the cut is at B-B. The sequential coordination required for walking requires the area below C-C to be operable. Simple tasks can be executed if the region below D-D is intact. Lengthy tasks and complex goals require the cerebral cortex (from /55/).

There are a number of different schemes existing for building up a hierarchical structure - see e. g. /56/ -. In general the lower levels try to reach some local targets on the basis of very detailed information, since the higher levels are influencing them to reach some global goals based on condensed information. For a technical example see e. g. /57/.

To imitate these ideas on the macrointelligence level with the hope to extend the power of such a device to handle unknown environment one has to solve the following two problems:

The first problem concerns the question of an adequate hierarchical structure including especially some guideline how to distribute the work load onto the different levels in this structure. A very interesting approach regarding this task has been given by G. N. Saridis in /58/: He establishes a connection between the entropy concept and feedback performance criteria for different levels in the hierarchy. By minimizing the overall information entropy, he gets an optimal apportionment of the overall task for the hierarchical system.

The second problem concerns learning in a hierarchy. Some investigations of this problem have been performed already in the area of stochastic automata (see e. g. /59/) and we have also considered such questions in connection with our microintelligence and macrointelligence structures (/60/). An important point seems to be, that it has to be expected that parallel learning is possible mainly in levels with very different, preferredly disconnected tasks such as abstract reasoning and senso-motoric process handling. For strongly connected tasks the lower levels of the hierarchy should have reached a good training status before learning for the higher levels is started. This is in accordance with the investigations on learning of an infant; the different levels of motion coordination and the different higher levels of intelligence are achieved only in a step by step procedure - see e. g. /7/ -. However, some counter experiences exist for a relatively simple example in which parallel learning was possible also for the same goal in two levels of hierarchy (/60/, /61/).

I.7. Recapitulation

In this introductory chapter, we have first discussed the general direction and goal of the work to be treated in detail in the following chapters. The expressions "microintelligence" and "macrointelligence" have been motivated and explained in their general meaning. The learning control loop being a combination of these ideas has been described and its behaviour demonstrated with the help of an example. Through a comparison with the well-known parameter-adaptive control concept, it was attempted to deepen the understanding of the learning control loop principle and its new aspects. Finally an attempt was made to give some insight into related work with a concentration on the major lines as seen by the author and illustrated by using selected papers/books rather than an exhaustive review.

I.8. Literature

/1/   Kroth, E.: Selbsteinstellendes Gleichstromantriebs-Regelsystem mit Mikrorechner. Darmstädter Dissertation 1985.

/2/   Geiger, G.: Fault Identification of a Motor Pump System using Parameter Estimation and Pattern Classification. 9th IFAC World Congress Budapest 1984

/3/   Isermann, R.: Digital Control Systems. Springer Verlag 1981

/4/   Schmidt, R. F. ed.: Grundriß der Neurophysiologie. Heidelberger Taschenbücher 4. Auflage. Springer Verlag 1979

/5/   Eccles, J. C.: The Understanding of the Brain. McGraw-Hill 1973

/6/   Nelson, P.G.: Neuronal Cell Cultures and their Application to the Problem of Plasticity in "Neural Mechanisms of Learning and Memory", ed. M. R. Rosenzweig and E. L. Bennett. MIT Press 1976

/7/   Piaget, J.: Psychologie der Intelligenz. 4. Aufl. Rascher Verlag 1970 8. Aufl. Verlag Clett-Cotta 1984

/8/   -: 8th International Joint Conference on Artificial Intelligence (IJCAI 83) Proceedings, Karlsruhe 1983

/9/   Newell, A.; Simon, H. A.: Human Problem Solving, H. A. Prentice Hall, Inc. 1972

/10/   Dörner, D.: Die kognitive Organisation beim Problemlösen. Verlag H. Huber 1984

/11/   Dörner, D.: Problemlösen als Informationsverarbeitung. Kohlhammer Verlag 2. Auflage 1979

/12/   Ersü, E.: Ein lernendes Regelungskonzept mit assoziativen Modellen von Neuronennetzwerken - oral presentation, no written version available - Regelungstechnisches Colloquium, Boppard 1983

/13/   Ersü, E.: On the Application of Associative Neural Network Models to Technical Control Problems in "Localization and Orientation in Biology and Engineering", ed. by Varju/Schnitzler" Springer Verlag 1984

/14/   Albus, J. S.: A New Approach to Manipulator Control: The Cerebellar Model Articulation Controller (CMAC) Trans. ASME 1975

/15/   Horn, B.K.P.; Raibert, M. H.: Configuration Space Control. MIT-AI Memo No. 458 1977

/16/   Raibert, M. H.: Analytical Equations vs. Table Look-up for Manipulation: A Unifying Concept. IEEE 8th Conference on Decision and Control, New Orleans 1977

/17/   Raibert, M. H.: A Model for Sensorimotor Control and Learning. Biol. Cybernetics 1978

/18/   Luh, Y. S.; Walker, M. W.; Paul, R.P.C.: On-line Computational Scheme for Mechanical Manipulators. J. Dyn. System. Meas. Control 1980

/19/   Ersü, E.; Rathgeber, K.; Schnell, M.; Neddermeyer, W.: A Robot Arithmetic Processor Concept for Cartesian Close-Loop Control with Prescribed Dynamics. Proc. IFAC Symposium SYROCO Barcelona, Spain 1985

/20/   Rojek, P.: Schnelle Koordinatentransformation und Führungsgrößenerzeugung für bahngeführte Industrieroboter. Tagungsband GMR, Fachtagung Steuerung und Regelung von Robotern - Langen, Germany 1986

/21/   Martinez, T.; Ritter, H.; Schulten, K.: Learning of Visuomotoric-Coordination of a Robot Arm with Redundant Degrees of Freedom in: Eckmiller, R.; Hartmann, G.; Hauske, G. ed. - Parallel Processing in Neural Systems and Computers. North-Holland 1990

/22/   Eckmiller, R.: Neural Computers for Motor Control in: Eckmiller, R. ed. - Advanced Neural Computers. North-Holland 1990

/23/   Arbib, M. A.; Kilner, W. L.; Spinelli, D. N.: Neural Models and Memory in "Neural Mechanics of Learning and Memory", ed. M. R. Rosenzweig, E. L. Bennett. MIT-Press 1974

/24/   Anderson, J. A.; Rosenfeld, E. (ed.): Neurocomputing - Foundations of Research MIT-Press 1987

/25/   Hebb, D.O.: The Organization Behaviour. John Wiley + Sons 1949

/26/   Rosenblatt, F.: The Perceptron, A Perceiving and Recognizing Automaton. Cornell Aeronautical Laboratory Report No. 85-460-1, 1957

/27/   Rosenblatt, F.: Principles of Neurodynamics, Perceptrons and the Theory of Brain Mechanisms. Spartan Books 1962

/28/ Minsky, M.; Paper, S.: Perceptrons - An Introduction to Computational Geometry, MIT Press 1969

/29/ Willwacher, G.: Fähigkeiten eines assoziativen Speichersystems im Vergleich zu Gehirnfunktionen. Biol. Cybernetics Vol. 24 1976

/30/ Ersü, E. : A Learning Mechanism for an Associative Storage System. Proc. IEEE Int. Conf. on Cybernetics and Society, Atlanta, USA, 1981

/31/ Marr, D.: Vision. Freeman Verlag 1982

/32/ Glünder, H. : On Functional Concepts for the Explanation of Visual Pattern Recognition Human Neurobiol. 1986

/33/ von Seelen, W.; Shaw, G.; Leinhos, U. M. (ed.): Organization of Neural Networks, Structures and Models. VCH- Verlagsgesellschaft 1988

/34/ Freeman, W. J.: An Epignetic Landscape of Brain Theory - see /33/ - 1988

/35/ Kohonen, T.: Content- Addressable Memories. Springer Verlag 1980

/36/ Marr, D.: A Theory of Cerebellar Cortex. Journ. of Physiology 202, 1969

/37/ Marr, D.: A Theory for Cerebral Cortex. Proceed. Royal Society London B176, 1970

/38/ Albus, J. S.: Theoretical and Experimental Aspects of a Cerebellar Model. Ph.D. Thesis, Univ. of Maryland 1972

/39/ Albus, J. S.: Data Storage in the Cerebellar Model Articulation Controller (CMAC) Trans. of the ASME 1975

/40/ Rosenzweig, M.R.; Bennett, E. L.: Neural Mechanisms of Learning and Memory. MIT Press 1974

/41/ Ito, M.; Sakurai, M.; Tongroach, P.: Climbing Fibre Induced Depression of Both Mossy Fibre Responsiveness and Glutamate Sensitivity of Cerebellar Purkinje Cells. J. Physiol. 234. 1982

/42/ Zhou, K. J.; Bai, J. K.: An Intelligent Controller of Novel Design. Proc. Control 85, Cambridge/UK 1985

/43/ Zadek, L. A.: Fuzzy Sets. Inf. Control 8. 1965

/44/ Zadek, L. A.: Outline of a New Approach to the Analysis of Complex Systems and Decision Processes. IEEE Trans. Syst. Man Cybern. 1973

/45/ Zimmermann, J.: Fuzzy Set Theory + and its Applications. Kluwer, Boston 1985

/46/ King, R. E.: Fuzzy Logic Control of a Cement Kiln Precalciner Flash Furnac Proc. IEEE Conf. on Applications of Adaptive and Multivariable Control. Hull, UK 1982

/47/ Yamazaki, T.; Mamdani, E. H.: On the Performance of a Rule- Based. Self- Organizing Controller. Proc. IEEE Conf. on Application of Adaptive and Multivariable Control Hull, UK 1982

/48/ Åström, K. J.; Anton, J.: Expert Control. Proc. 9th IFAC World Congress Budapest, Hungary 1984

/49/ Bieker, B.: Wissenserwerb für eine einfache Expertensystem- Regelung atp 1986

/50/ Saridis, G. N.: Self- Organizing Control of Stochastic Systems. Marcel Dekker 1977

/51/ Saridis, G. N.; ; Mendel, J. M.; Nicolic, Z. J.: Report on Definitions on S.O.C. Processes and Learning Systems. IFAC Trans. Syst. Man Cybern. 1974

/52/ Narendra, K.S.: Recent Developments in Learning Automata Theory Applications. Proc. 3rd Yale Workshop on Applications of Adaptive Systems Theory 1983

/53/ Narendra, K.S.; Mars, Ph.: The Use of Learning Algorithms in Telephone Traffic Routing - A Methodology. Automatica 1983

/54/ Hämäläinen, R.P.: On a Class of System Models in Physiology, Acta Polytechnica Scandinavia Mathematics and Computer Science, Series 31 1979

/55/ Albus, J.S.: A Model of the Brain for Robot Control - Part 3: A Comparison of the Brain and our Model Byte 1979

/56/ Findeisen, W.; Brdys, M.; Malinowski, K.; Tatjewski, P.: Control and Coordination in Hierarchical Systems. J. Wiley 1980

/57/ Reinisch, K.: Hierarchical On-Line Control of Disturbed Dynamic Processes. Proc. 4th IFAC/IROS Symp. Large Scale Systems: Theory and Applications. Zürich, Switzerland 1986

/58/ Saridis, G.N.: An Integrated Theory of Intelligent Machines by Expressing the Control Performance as Entropy. Control-Theory and Advanced Technology 1985

/59/ Baba, N.: Learning Behaviours of Hierarchical Structure Stochastic Automata Operation in a General Multiteacher Environment. IEEE Trans. Syst. Man Cybern. 1985

/60/ Ersü, E.; Tolle, H.: Vertiefende Untersuchungen zu gemäß der menschlichen nervösen Informationsverarbeitung lernenden Systemen. Abschlußbericht für das von der VW-Stiftung geförderte Vorhaben I/60/901 TH Darmstadt 1988

/61/ Ersü, E.; Tolle, H.: Hierarchical Learning Control - An Approach with Neuron-like Associative Memories in: Anderson, D. ed. - Collected Papers of the IEEE Conf. on Neural Information Processing Systems, Denver (USA), 1987 - American Institute of Physics 1988

II. Microintelligence

II.1. The cerebellar model of J. S. Albus

II.1.1. Neurophysiological background

As has been discussed in chapter I, the microintelligence level comprises the effective storage and retrieval of useful situation-dependent information. The starting point in our research on technical applications of various devices has been a cerebellar model as put forward by J. S. Albus in /1/, /2/. A short description of this model and its neuronal background is given in the following paragraphs. However, the discussion of this background is for information only: it should be remembered that most models of reality used for engineering purposes are composed of only those features, which seem to be relevant for a particular application. Therefore it is of no importance, whether the Albus model describes the details of human brain functioning correctly or not as long as it produces satisfactory results in the learning control loops studied. (Indeed some completely different interpretation for the purpose and operation of the cerebellum has been given by A. J. Pellionisz - see e. g. /3/ -).

Fig. 1 is a schematic sketch of the human brain, in which some of the most important areas for sensomotoric reactions are marked. The cerebellum, which is discussed here, is intimately involved with the control of rapid, precise, coordinated movements of hands, limbs and eyes as one can conclude from cases of injuries to the cerebellum. It exhibits a fairly regular structure as can be seen in the three dimensional diagram of fig. 2a due to Eccles et. al. (/5/). The simplified extraction given in fig. 2b (from /4/) is used in the following rough discussion of the possible functions of the different elements shown.

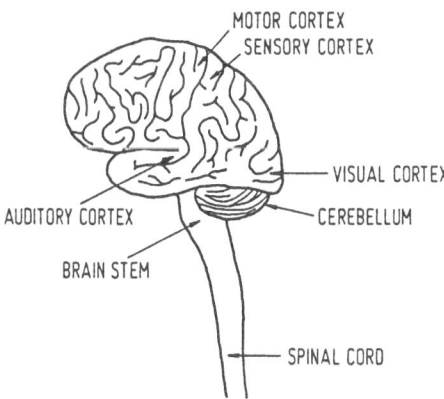

Fig. 1 (from /4/): Side view of the human brain

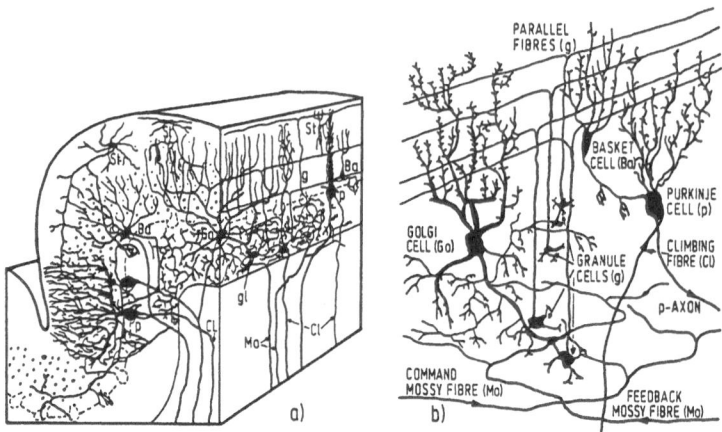

Fig. 2 (from /5/, /4/): Cerebellum with inhibitory Golgi, Basket and Stellate Cells (Stellate cells not shown in 2b), incoming Climbing and Mossy Fibres, Granule Cells imbedded in the cerebellar glomeruli (gl in 2a, not shown in 2b) with the Parallel Fibres as their axons and Purkinje Cells, which axons are the only outgoing fibres.

The mossy fibres are the inputs to the cerebellum. They combine through different strings information about the current status and situation of the individual through sensory feedback and also about the envisaged goals by signals from higher levels of the brain. Each of the mossy fibres makes contact with several hundreds from the 10^9 to 10^{10} granule cells in the cerebellum. Furthermore some overlap exists, since each granule cell is contacted by 5-10 mossy fibres. The axons of the granule cells, the T-shaped lines of fig. 2b, are passing on this mixed incoming information to the stellate, basket and Purkinje cells.

The overall granule cell activity is controlled by Golgi cells to a fairly constant level of activity of 1 % or even less of all granule cells. For this purpose the Golgi cells sample information through their dentritic trees from the mossy fibres as well as the granule cell axons and then make inhibitory contact through their branching axons with up to 100 000 granule cells. So the incoming information from the sensoric evaluation of the situation and the resulting decisions of the cerebrum are carefully combined in the sense that a constant amount of information, selected by the level of exitation, is delivered to the next stage of information processing.

This next stage is a weighted adding up of the parallel fibre activity through the stellate, basket and Purkinje cells. The stellate and basket cells, for which the major distinction (besides that of the cell geometry) is their location in the molecular layer, seem both to perform the same function of inverting the excitation from the parallel fibres into inhibitory signals, which they transmit to the Purkinje cells. Since the Purkinje cells add up their inputs without inversion, this mixture of plus and minus signs produces a much increased ability (roughly by a factor of 1000) for input pattern discrimination than would exist otherwise. For details see /1/.

The weights, by which the activity of the parallel fibres is evaluated, are modifiable. The modification is achieved through signals transmitted by the climbing fibres, where each one typically works together with just one Purkinje cell. This fibre climbs over the dendritic extensions of the Purkinje cell like ivy on a tree. The climbing fibres stretch further some collaterals to the soma of adjacent basket and stellate cells, being able in this way to influence the level of the respective inhibition. The information distributed by the climbing fibres seems to be sensoric information, however, having had a careful preprocessing by other parts of the brain. So the climbing fibre signals may represent an evaluation of the amount of success or failure regarding a desired action.

The only output of the cerebellum is finally produced by the axons of the Purkinje cells, which are transmitting the result of the summing of the excitatory signals from themselves and the inhibitory signals from the stellate and basket cells.

II.1.2. Translation of the neurophysiological findings into a model implementable on a computer

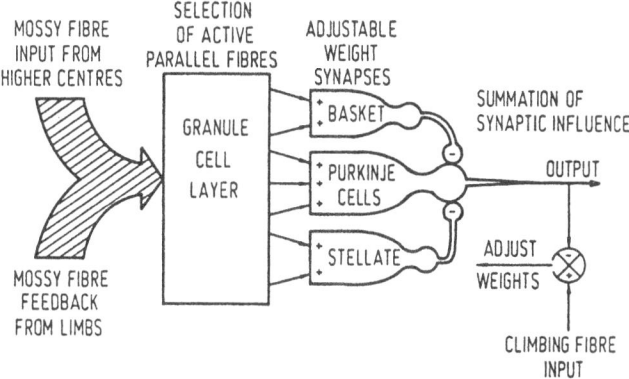

Fig. 3 from /6/: General scheme of responsibilities and organisation in the cerebellum

Fig. 3 from /6/ summarizes the neurophyiological ideas about the functioning of the cerebellum in form of a schematic structural sketch. As a first step in the direction of a computer model it is helpful to discuss the differences between sensory feedback input and inputs from the higher centres of the brain. The sensory feedback inputs will be continuous signals within a certain range. The inputs from higher centres are either discrete signals like "flee" or "attack" or a mixture of discrete and continuous signals like "move" with a certain speed out of a possible range. One can now handle continuous and discrete signals either in a uniform or in a non-uniform way. In the first case, one may call all mossy fibre inputs s_i, specify for each s_i a range $s_{imin} \leq s_i \leq s_{imax}$ and so one would have for continuous signals the possibility of continuous change between s_{imin} and s_{imax}, for discrete signals e.g. a two level situation, in which s_i can take on the values $s_i = s_{imin}$ or $s_i = s_{imax}$ only. In case of a non-uniform treatment, one may use the discrete values to characterize a certain set up or "context", for which the continuously changing values s_i are then the remaining input variables. In his original work /1/, J.S. Albus used a non-uniform treatment, as indicated in the following description. Later on, he turned to the uniform description - see /2/ -, which is also the version studied in detail by us, especially since in control loops one has normally to deal only with continuous signals.

Fig. 4 from /1/ gives some details on the non-uniform treatment of discrete signals describing the general requirements through a "task name" as "move" and of continuous signals as "position and velocity measurements". We shall start with a discussion of the handling of the continuous signals, for which the idea of local generalization is basic. The right-hand side of fig. 4 showing the selection of active synaptic weights and their adjustment ist explained also. The term "hash coding", its use for task selection and the advantages of such codes will be commented on then.

The translation of the scheme of fig. 3 into a model suitable for a digital computer results, for the continuously changing variables and/or a uniform treatment of continuously changing and discrete variables, in a structure shown in fig. 5: The inputs coming from sensory feedback and/or higher levels of the brain are written uniformly as inputs $s_1, s_2 \dots s_n$.

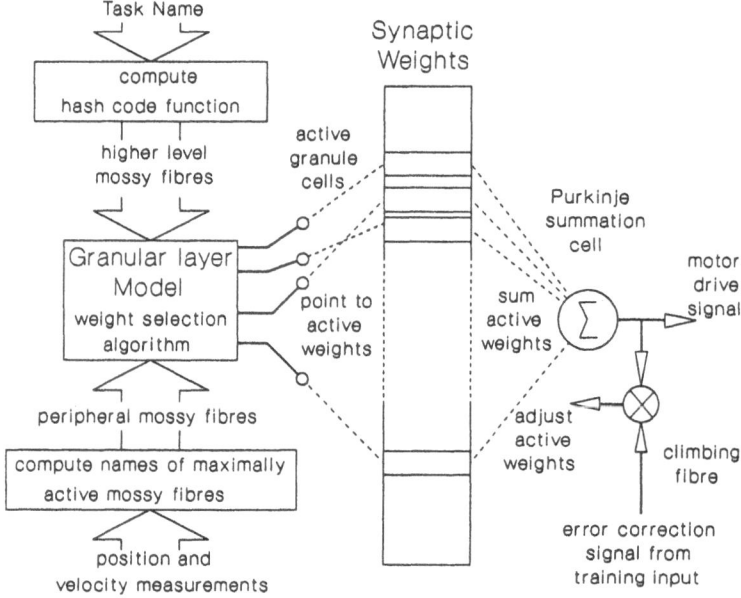

Fig. 4: More sophisticated sketch of the general scheme of fig. 3 regarding general inputs from higher levels of the brain and continuously variable sensory feedback information (from /1/)

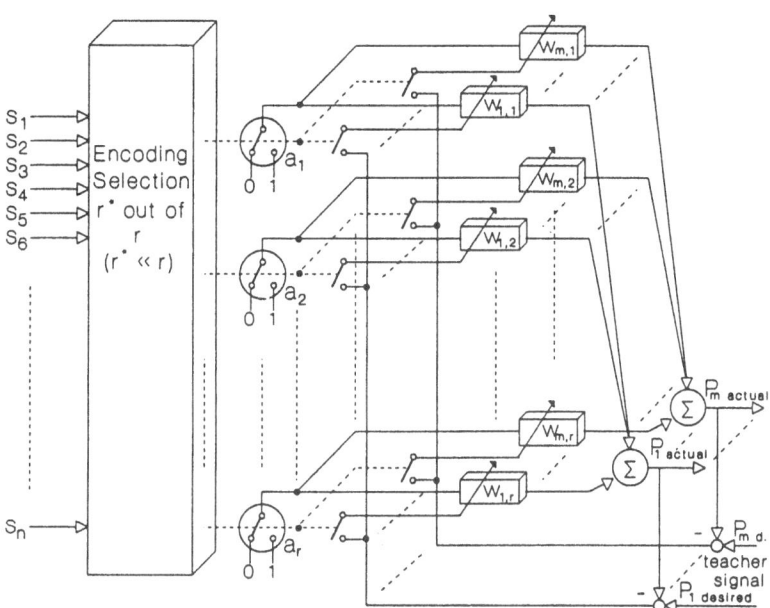

<u>Fig. 5 from /7/</u>: Model for computer implementation of the general scheme from fig. 3 - explanation in the text

The "Encoding", to be explained later on, distributes the incoming information for each of the eventual outputs onto a small numer r^* out of r available memory places for this output. This is characterized by the switches following the block "Encoding". The content of the memory places are "weights" w_{ij}, the first index being the number of the respective output, the second index indicating, which one of the r memory places is used for storage of this value. The output p_i is in the most simple case just the sum of the weights in the r^* memory places activated by the actual input. Since the stored values w_{ij} can be positive or negative in a computer, a distinction between excitatory Purkinje cells and inhibition by basket and stellate cells is not necessary in the computer model. The mechanism of memory training is indicated in fig. 5: The actual output for a certain input $s_1...s_n$ is compared with the desired output and the difference is used to change the weights in the active memory places in such a manner that the next time the same input is applied, the desired output is generated. This correction is made e. g. by adding $1/r^* \cdot (p_{desired} - p_{actual})$ at all r^* active memory places to the existing values of the weights stored there.

For the distribution of the incoming information onto r^* different places - the "Encoding" (with the exception of the hash coding procedure to be discussed later on) - J.S. Albus exploits some empirical findings by J.K.S. Jansen et. al.(/8/) that individual mossy fibres seem to fire at their maximal rate, when specific conditions exist in specific parts of the periphery. E.g. a mossy fibre carrying elbow position information will fire at its maximum rate, when the angle α describing the elbow position is between 40^0 and 80^0. A certain number of mossy fibres may so cover the whole range of possible positions 0^0- 160^0 by a fairly broad range of conditions: MF1, MF2, MF3, MF4 (see fig. 6). The necessary narrowing down of position descriptions to intervals useful for the human being is achieved by shifting the conditions for firing by a small amount without changing the broad range for the maximum firing rate (see again fig. 6).

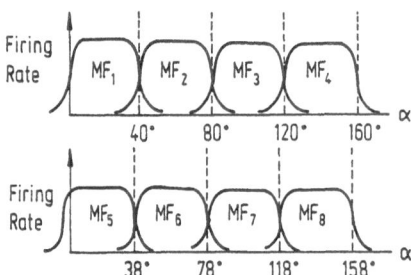

Fig. 6 from /1/: Two sets out of 20 assumed firing rate patterns, leading individually to a coverage of the whole range of possible elbow positions without much accuracy, but generating this accuracy - 2^0 - by being shifted against each other by this amount.

A mathematical description of this procedure has the following form - see fig. 7 -: Assume that we consider an output $p=f(s)$ on the interval $s_A \leq s \leq s_E$ by a digital computer. We have then to discretize the interval $s_E - s_A = L$, by deviding L into R subintervals of length ϵ with ϵ being the smallest amount of accurate description for s, due to the fact we have to represent s by a finite number of digits in a computer and/or that the measurement device is accurate only up to a certain number of digits. An equal description of points in the interval of length L is by dividing L into r^* interval-groups of length $r^* \bullet \epsilon$, shifted against each other by ϵ, and describing each individual point by the r^* intervals from the different groups, to which it belongs.

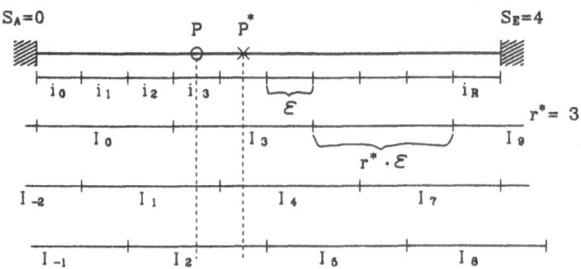

Fig. 7: P from the interval [0, L] may be described up to the accuracy ϵ either by being inside of the interval i_3 or by being inside of the interval-group I_3, I_1, I_2. In each case another point being in another element of the original ϵ-frame is described by some other naming e. g. p^* by i_4 or $I_3 I_4 I_2$.

The characterization by broader, shifted intervals leads, together with the adding up of information as performed by the Purkinje cells, to local generalization if for a certain interval not the actual output value connected with P but $1/r^*$ of this output is stored. One concludes this immediately from fig. 8, which shows this for the situation of fig. 7 and a certain assumed output for P. As we shall discuss later on, such a scheme of local generalization has not only the advantage of creating some first answer in the neighbourhood of a trained situation, when situations from this neighbourhood are met for the first time, but it reduces also the storage capacity required and - what seems even more important - the training effort necessary to get a sufficiently accurate output for the range of interesting inputs.

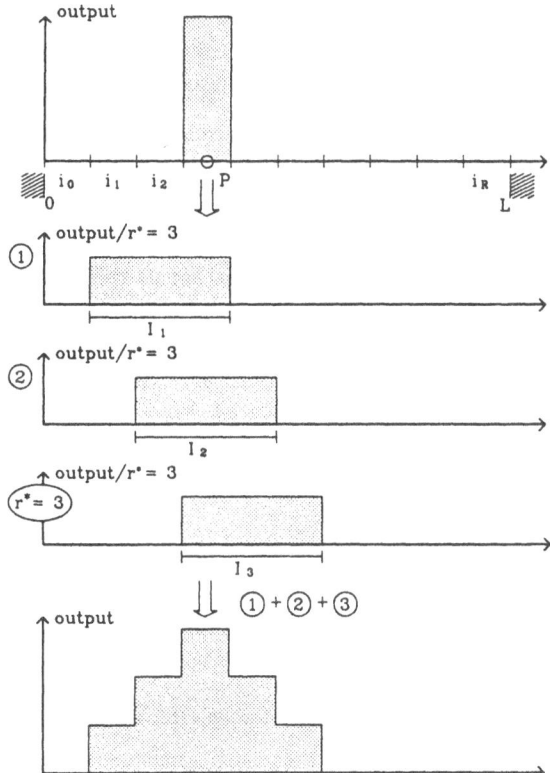

Fig. 8: Local generalization through adding up of $(1/r^* \bullet \text{output})$ stored in the broader intervals of length $\epsilon \bullet r^*$. Note, that any quantization by ϵ gives a certain generalization of the output. However, in the normal storage mode one gets an answer only for a stored value in its ϵ-neighbourhood but for the distributed storage one gets together with the correct answer in the ϵ-neighbourhood some similar - possibly not too wrong - answer in a $r^* \bullet \epsilon$ neighbourhood, the similarity being a function of the distance from the stored value.

We shall now turn to "hash-coding". Hash-coding is the most frequently used procedure for a fast access to memory places. The basic idea is to generate the address where certain data shall be stored by computing this address from the content of the data. One speaks therefore also of content addressable and/or associative memories. A detailed description of the respective background and methods is e.g. given by T. Kohonen in /9/. Since s is in our case in general a n-dimensional input-vector, for which the respective output has to be stored onto a physically two-dimensional memory, one has the problem to set up some procedure for mapping the n-dimensional input space onto the two-dimensional storage space, anyhow. So it makes sense instead of inventing some n-to-two dimensional storage place mapping for the possible inputs to use instead the situation representing input data as a keyword, out of which appropriate memory places are computed (hash-coded) especially since this is known to be a very effective way for fast memory access. Details of the procedure used by us will be discussed in section II.2.3. However, one further advantage of "hash-coding" for our problem should be mentioned already here. To store a certain input element s_i we have to define an input range $s_{iA} \leq s_i \leq s_{iE}$, an accuracy ϵ_i and a local generalization extension r^*. The product $r_i^* \epsilon_i$ determines the region of local generalization (compare fig. 8), the interval length $s_{iE} - s_{iA}$ together with ϵ_i the number of elements of s_i between which one has to differentiate (compare fig. 7) and especially the number of necessary memory places for the respective output. In general, one can only estimate the upper and lower limits for a certain input. This leads to safety margins and the real inputs will only cover part of the range, which is formally attributed to them, seemingly making necessary more memory space than is actually used later on. This situation becomes worse and worse with a growing number of inputs and is sketched for two dimensions in fig. 9.

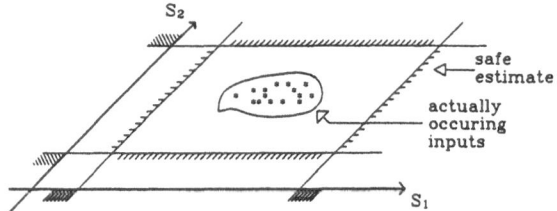

Fig. 9: Overestimation of necessary memory space through safe estimation of input ranges.

Now, hash-coding is again advantageous with respect to this problem. Since storage addresses are generated out of the content of input data, the necessary amount of storage places is not dependent of range estimates but on the amount of input data, for which certain values have to be stored. Since the number of possible different input-values being relevant for the control task is not known in advance, some estimate on the necessary physical memory size has to be made. The problems connected with this will be considered after a short description of how the hash-coding procedure works in general.

This can be described as follows. From the intervals activated by a certain input-value (as shown in fig. 7 for the one-dimensional case) a memory address is calculated by a pseudorandom algorithm. "Pseudorandom" means here that the addresses are chosen randomly from the available memory places, however in such a way, that if the same input comes up again, the same address of the memory will be computed again. Two problems can arise with this procedure: First, the random number generator may produce the same address for two different inputs. One must try to avoid this situation as far as possible by suitable measures which will be described together with our memory version. However some of these so-called "hash-collisions" will remain in spite of this. They can be tolerated, if their number is very small, since they have the same effect as an incorrect information transmission of some element of an information chain, against which the control algorithms have to be robust, anyhow. The second problem is an underestimation of the necessary size of the memory: This leads unavoidably to a large number of hash-collisions and has to be discovered during training before such a situation occurs by some bookkeeping on the number of different inputs. The only possible solution for this problem is an extension of the memory provided.

Hash-coding has, in addition to the reduction of the necessary physical memory the further advantage that the destruction of a certain part of the memory does not destroy a certain range of input information but only randomly distributed parts of the information to be stored. Since with local generalization the output value to be generated by a certain input is the sum of the content of r^* storage places, in general only part of the weights contributing to the output value are taken away through the memory destruction. That means the memory output for a given input will not be completely wrong, but only less exact, provided that the part of the memory which has been destroyed is not too large. Since such a behaviour is also a property of the brain, hash-coding used in the encoding procedure helps us to imitate further neurophysiological reality to some extent.

Up to now, we have not discussed the handling of discrete information.

If such information is taken into account by the s_i-inputs without certain precautions, local generalization would just destroy the sharp distinction between different situations characterized by discrete information. However the only precaution necessary is to model the discrete information as constant values

- being present over a range much larger than r^*, so that the extension at the border by local generalization is negligible

- and being separated by more than r^*, so that mixing up by local generalization is impossible.

Another way, which avoids the direct use of the s_i-inputs is to select different hash codes for different set-ups (contexts) described by the discrete information. In this case different memory places would be selected for the same continuous s_i-inputs by the different hash codes initiated through the different discrete task names. If the physical memory is big enough, this allows one to store different behaviour depending on the context and/or the envisioned task. In fig. 4 exactly this procedure is put forward on the left-hand side, representing the "encoding" and/or memory place selection.

These remarks conclude the general discussion of Albus' translation of neurophysiological knowledge and/or speculation about the cerebellum into a model, which can be implemented on a digital computer. In the subsequent more detailed considerations we shall concentrate on the mathematical formulations and software developed by us without making much further reference to this background.

II.2. AMS - a computer oriented effective implementation of the cerebellar model of J. S. Albus [1]

II.2.1. Introductory remarks

The general aim of our work, to apply neurophysiological ideas to real time control of technical processes, made it necessary to look from the beginning for a very time-effective, memory saving implementation of the microintelligence element.

Due to the fact that it is used for storing the relation between a certain stimulus/situation and response/reaction and that this is called in psychology an "association" we have called the result "AMS" = "Associative Memory System".

In the following section the layout of the AMS will be explained as far as this is possible without making reference to the special computer and/or computer language used. Actually, for the real implementation it turned out to be necessary to leave for certain aspects the more general description of the system as it is represented by a program in a high level language (for instance FORTRAN) and to make use of certain procedures available in assembler language only. These details will not be treated, since they are too much dependent on the hard- and software used.

Mathematically the AMS results in an overall mapping H from n inputs s_i onto m outputs p_j:

(1) $H: \underline{s} \rightarrow \underline{p}$; $\underline{s} \in \mathbb{R}^n, \underline{p} \in \mathbb{R}^m$

---

[1] A carefully tested commercial version of AMS designed for the application with modern computers is available from ISRA Systemtechnik GmbH, 6100 Darmstadt, Mornewegstr. 45A, Germany.

This mapping can be subdivided into a series of mappings:

(2a) $h_1: \underline{s} \rightarrow M$

(2b) $h_2: M \rightarrow A$

(2c) $h_3: A \rightarrow \underline{p}$

h_1 determines out of the stimulus content a Matrix M, which is a $(n \times r^*)$-matrix computed from the r^* active intervals in each dimension of $\underline{s}$.

h_2 is then responsible for the selection of the memory addresses, which is accomplished by building a new row out the n rows of M and using the content of this new row to generate the respective address with some hash-coding procedure. The necessary different addresses for the m outputs could be generated either by m different hash-codes, which may be difficult to produce, or more easily by some deterministic procedure like adding $(i-1) \bullet r$ to the address found for storing p_i. It would be then an rxm Matrix, r being the number of memory places foreseen for each of the outputs. By this means one would get memory locations for all outputs with one hash-coding procedure. However, the stimuli-scalings/generalisations and the output subdivision would be the same for all outputs. This is a bad solution if one has a task, for example in which the stimulus element s_i is of high significance to the output p_j but of low significance for the output p_k. Unnecessary resolution reached by wrong scaling/generalisation does mean, however, unnecessary learning effort. There-fore, it is in general better to use instead of one AMS with m outputs m AMS-elements with one output, leading to $A = \underline{a}$, a vector with r elements. One pays for this flexibility by generating the hash-code m times instead of once, which takes up computing time in case of serial computation but could be handled in principle without this disadvantage by parallel processing (i.e. use of m processors). For this reason we have concentrated ourselves on this latter solution. So in the following we always consider the AMS device with one output only.

h_3 is finally the procedure to sum the contents of the active cells from the vectors $\underline{a}_i$ to give the output values p_i. Different possibilities exist to perform this in detail and a number of them considered by us will be discussed.

The description of AMS is based on /10/ and up to now unpublished material. It is divided in accordance with the subdivision of the overall mapping H into the three submappings h_1-h_3.

II.2.2. The first mapping: Numerical input value characterization

If one does not consider the development of special hardware - which is an approach, which we just consider in a research project supported by the German Science Foundation/DFG - one has to take into account the general working conditions of current computers to reach an efficient realization of the neuronal network imitation.

The first decision for an algorithmic implementation on a computer concerns the representation of numbers. Two possibilities exist: floating point and fixed point representation. In the fixed point representation only whole numbers of a certain interval, e.g. - 2^{15} till 2^{15}-1 (- 32768 till 32767) are possible. The advantage is, that such numbers need only very little memory space (2Bytes) and can be handled very simply and fast. The floating point representation allows much bigger intervals to be covered, e.g. - $1,7 \cdot 10^{38}$ till +$1,7 \cdot 10^{38}$ with a resolution of 10^{-7} for each value. The large interval and the high resolution have to be payed for by twice the memory places with respect to the fixed point numbers and by a much more complicated handling of the numbers, so that without a special hardware - a floating point processor - all arithmetic steps are much slower as in the fixed point case.

The transition between fixed point and floating point representation requires a certain amount of conversion work. To avoid unnecessary computations, all values to be handled in the AMS, the inputs, the outputs and the intermediate representations of the outputs in r^* memory cells have to be of the same representation type.

The fixed point representation was selected for the following reasons: The resolution provided by the fixed point representation seemed to be sufficient in connection with the quantization and local generalization used anyhow. However, computation speed is a very dominant requirement for handling real processes and for this the fixed point representation is much better as pointed out already. Finally, the analog-digital- and digital-analog-converters being necessary for coupling computers with technical processes work with the fixed point presentation also. The only disadvantage connected with the selection is the necessity of careful scaling for the inputs, to make best use of the available interval for fixed point numbers.

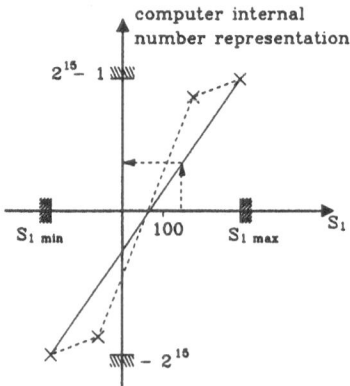

Fig. 10: Example for scaling; s_{1max} = 300; s_{1min} = - 150 is projected onto the fixed point number representation -2^{15} till 2^{15}-1 by the continuous line in a linear way, by the broken lines in a nonlinear way, giving in this latter case a higher resolution to the center area of the variation of the stimulus element s_1

Fig. 10 indicates, that such a scaling might be achieved by a fixed factor b_i in a linear way or by a nonlinear function $b_i(s_i)$ - in this case constructed out of three linear parts - to improve resolution in certain important areas.

A nonlinear scaling may be wanted e.g. due to the fact, that one expects the really interesting stimuli in a certain region of the stimulus interval only, the larger stretching of the interval being a safety measure, or even necessary, since the resolution in the fixed point number area is limited to $\epsilon_{max} \approx 1/2 \bullet 2^{15} \approx 0,0015\ \%$, so that the required stimulus resolution:

$$(3) \qquad \epsilon_i = \frac{|s_{imax} - s_{imin}|}{R_i} \qquad\qquad R_i = \text{quantization thought to be neccessary.}$$

cannot be reached for the whole stimulus interval. However, the latter case will be very rare in technical applications.

The intervals produced in the whole number region from -32768 till +32767 by R_i are originally limited by non-whole numbers. So the first step is a rounding off. But to improve the handling of the intervals the length of those intervals is reduced in a second scaling down to a value of one, so that for all further steps all stimuli have a basic resolution of $\tilde{\epsilon}_i$ = 1 for all i.

To discuss the internal structures and properties of AMS further, we shall use the example of two input stimuli s_1, s_2, which may be, for example, the input u(k) and the output y(k) of a predictive process model without history in the inputs and with the response p = y(k+1) (compare section I.4.). The local generalization as discussed in connection with fig. 6 and fig. 7 leads now to intervals of the length r^*, shifted by $\tilde{\epsilon}$ = 1. Due to the scaling done before, it is not really necessary - although possible in general - to deal with different generalization factors for the different stimuli s_i. Fig. 11 shows the original $\tilde{\epsilon}$ = 1 resolution, the shifted intervals of length r^* for r^* = 4, the resulting coarser grids and for three different s_1,s_2-combinations the local generalization being generated by superimposing the active areas of the coarser grids.

The local generalization areas are not equal in their form, but this levels out in the AMS output, if one considers a large number of points, as being normally the case for learning elements. Suggestions for improving the uniformity of the generalisation areas have been put forward in Parks and Militzer /17/.

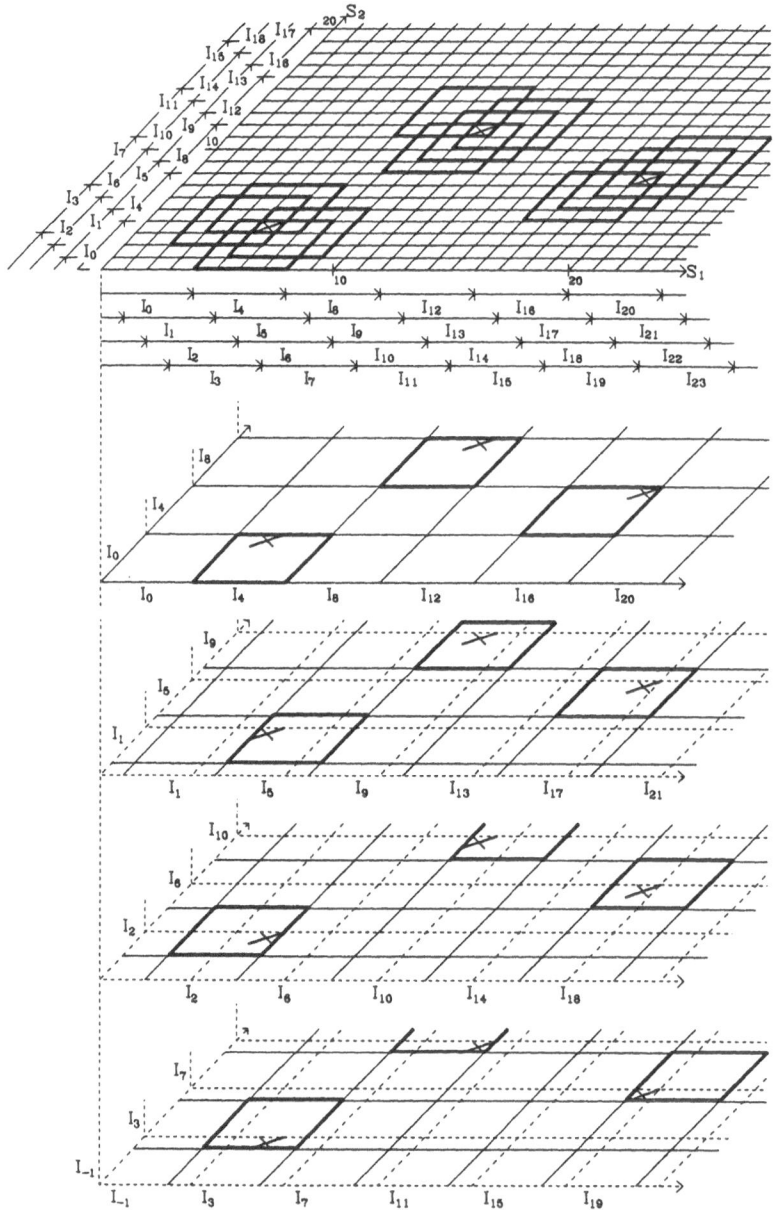

<u>Fig. 11:</u> Local generalization for two input stimuli s_1, s_2 and $r^* = 4$ with marked shifted intervals of length r^* and respective coarser grids. Points: $(s_1, s_2) = (5,4; 3,4)$; $(s_1, s_2) = (14,4; 6,4)$; $(s_1, s_2) = (7,4; 12,4)$. One can see the local generalization achieved automatically by the shifted intervals, however too, that the generalization areas are not of fully equal form.

In fig. 11 points were marked with non-integer values. This was only done to obtain a better picture: in AMS all values are rounded to whole numbers. The intervals, being characterized by their lower limit, are starting with the respective whole number κ and are ending with $\kappa + r^* - 1$. Each coarser grid is characterized by intervals with the same division rest from the division of a whole number by r^*. Therefore the division rest λ can be used as an indicator for the grid considered. The active intervals can be memorized by the factors ρ, with which r^* has to be multiplied to reach together with a division rest smaller than r^* the whole number just under inspection, and by storing this factor at a place reserved for the coarser grid characterized by the respective division rest λ. Using the definition, that the intervals contain the integer at which they start and which is characterizing them as an index in fig. 11, and just do not contain the integer at which they end, one gets for the three points in fig. 11 the following characterization:

a) $P_1 = (5,4; 3,4) \xrightarrow{\text{rounded}} (5;3)$.

$$5 \notin I_6 \mathbin{\hat{=}} (\rho=1)\bullet(r^*=4)+(\lambda=2) \Rightarrow \begin{cases} \in I_5; & I_4; & I_3; & I_2 \\ \rho=1 & \rho=1 & \rho=0 & \rho=0 \\ \lambda=1 & \lambda=0 & \lambda=3 & \lambda=2 \end{cases}$$

$$3 \notin I_4 \mathbin{\hat{=}} (\rho=1)\bullet(r^*=4)+(\lambda=0) \Rightarrow \begin{cases} \in I_3; & I_2; & I_1; & I_0 \\ \rho=0 & \rho=0 & \rho=0 & \rho=0 \\ \lambda=3 & \lambda=2 & \lambda=1 & \lambda=0 \end{cases}$$

b) $P_2 = (14,4; 6,4) \xrightarrow{\text{rounded}} (14;6)$.

$$14 \notin I_{15} \mathbin{\hat{=}} (\rho=3)\bullet(r^*=4)+(\lambda=3) \Rightarrow \begin{cases} \in I_{14}; & I_{13}; & I_{12}; & I_{11} \\ \rho=3 & \rho=3 & \rho=3 & \rho=2 \\ \lambda=2 & \lambda=1 & \lambda=0 & \lambda=3 \end{cases}$$

$$6 \notin I_7 \mathbin{\hat{=}} (\rho=1)\bullet(r^*=4)+(\lambda=3) \Rightarrow \begin{cases} \in I_6; & I_5; & I_4; & I_3 \\ \rho=1 & \rho=1 & \rho=1 & \rho=0 \\ \lambda=2 & \lambda=1 & \lambda=0 & \lambda=3 \end{cases}$$

c) $P_3 = (7,4; 12,4) \xrightarrow{\text{rounded}} (7;12)$

$$7 \notin I_8 \mathbin{\hat{=}} (\rho=2)\bullet(r^*=4)+(\lambda=0) \Rightarrow \begin{cases} \in I_7; & I_6; & I_5; & I_4 \\ \rho=1 & \rho=1 & \rho=1 & \rho=1 \\ \lambda=3 & \lambda=2 & \lambda=1 & \lambda=0 \end{cases}$$

$$12 \notin I_{13} \mathbin{\hat{=}} (\rho=3)\bullet(r^*=4)+(\lambda=1) \Rightarrow \begin{cases} \in I_{12}; & I_{11}; & I_{10}; & I_9 \\ \rho=3 & \rho=2 & \rho=2 & \rho=2 \\ \lambda=0 & \lambda=3 & \lambda=2 & \lambda=1 \end{cases}$$

These characterizations can be compressed into a matrix - which we shall call M-matrix -, in which one uses for each of the n stimuli one row - that means the first row for stimulus s_1, the second row for stimulus s_2 and so on - and for each division rest, starting with $\lambda = 0$, one column, so that one gets a $n \times r^*$-matrix, the elements of the M-matrix being the multiplication factors ρ. For our example points means that

$$
\begin{array}{cccc}
& \lambda=0 & \lambda=1 & \lambda=2 & \lambda=3 \\
\end{array}
$$

$$
M_{5,4;3,4} = \begin{bmatrix} 1 & 1 & 0 & 0 \\ 0 & 0 & 0 & 0 \end{bmatrix}
$$

$$
M_{14,4;6,4} = \begin{bmatrix} 3 & 3 & 3 & 2 \\ 1 & 1 & 1 & 0 \end{bmatrix}
$$

$$
M_{7,4;12,4} = \begin{bmatrix} 1 & 1 & 1 & 1 \\ 3 & 2 & 2 & 2 \end{bmatrix}
$$

One sees easily by inspection of the examples, that the M-matrix elements can be generated very quickly from the components of the stimuli vector in the following way: after rounding a component, one divides it by r^* and stores the integer multiplication factor under the resulting integer division remainder. All elements to be left of this place in the same row have the same value, all elements to the right of this place have this value minus one, so that they can be filled in automatically. For negative numbers one considers instead of the negative integer division remainder λ the value $\lambda + r^*$, which gives the positive row number and one stores instead of the integer multiplication factor in the left area this negative value plus -1 and in the right area now the original negative value. This modification takes into account the fact that the intervals are stretching from their index value to the right as well in the region of positive numbers and in the region of negative numbers.

The general formula for generating the elements m_{ij} of the M-matrix is given by:

(4a) $\chi_i = $ rounded $s_i \overset{!}{=} \rho \cdot r^* + \lambda \quad \rho, \lambda = $ integers

$$
m_{ij} = \begin{cases} \chi_i \geq 0 & \begin{cases} \rho & j \leq \lambda \\ \rho - 1 & j > \lambda \end{cases} \\ \chi_i < 0 & \begin{cases} \rho - 1 & j \leq \lambda + r^* \\ \rho & j > \lambda + r^* \end{cases} \end{cases}
$$

One can state finally that by using integers only and by normalizing the quantization to the value $\tilde{\epsilon} = 1$ for all stimuli one gets with the scheme discussed a very fast computational method for generating the first mapping h_1, the starting point for the content-driven hash-coding memory place selection, to be discussed in the next section.

II.2.3. The second mapping: From matrices to memory locations

We have now to deal with the problem of memory place address generation from the content of the M-matrices. Actually, we applied certain bit-manipulation procedures already available in the software of the DEC machine used by us to accelerate the pseudo-random hash-coding as much as possible. These machine specific possibilities will not be discussed in detail, but we shall restrict the following description to the general ideas taken into account.

In a first step a translation is made from the original values in the M-matrix into a hexadecimal representation of figures as shown by the following example ($\underline{s}^T = (s_1, s_2, s_3, s_4)$, $r^* = 4$):

$$\begin{bmatrix} s_1 \\ s_2 \\ s_3 \\ s_4 \end{bmatrix} \rightarrow M = \begin{bmatrix} 1 & 1 & 1 & 0 \\ 2 & 2 & 2 & 2 \\ -4 & -5 & -5 & -5 \\ 3 & 3 & 2 & 2 \end{bmatrix} \xrightarrow[\text{representation}]{\text{hexadecimal}} \begin{bmatrix} 01 & 01 & 01 & 00 \\ 02 & 02 & 02 & 02 \\ FC & FB & FB & FB \\ 03 & 03 & 02 & 02 \end{bmatrix}$$

The elements of the hexadecimal representation are then assembled together for each column n to give a subelement of a row vector representing M:

$$\rightarrow (\text{03FC0201} \quad \text{03FB0201} \quad \text{02FB0201} \quad \text{02FB0200})$$

that means each subelement represents the values of the stimuli, which give the same division rest from the division by r^*. This leads to r^* subelements in the row vector.

All subelements like 03FC0201 of this row vector are different by construction with the result that they will point to a different storage place address when their content is used for a calculation of such an address by hash-coding. The number of different addresses is in agreement with the wanted distribution of the input data onto r^* different places. So we have reached an adequate situation for calculating the actual storage place addresses, if we use for this the subelements.

Now, the length of the subelements is in general dependent on the number of inputs. To avoid this, four inputs are always clustered into one subelement, giving fixed zeros in the subelements if we have less than four inputs and leading to more than one row vector, if there are more than four inputs. Using for different row vectors different hash-coding algorithm and generating the needed address out of some overlay of hash-coded values from the different rows, one can easily live with this clustering of inputs. The advantage is, that each subelement is fixed now onto a 32 bit word. So the hash-coding can be concentrated on a pseudo-random processing of the content of 32 bit words.

Actually, the storage place address is generated from 16 bits of the 32 bit word only. The other 16 bits are used with eight bits for building up a so-called identifier which helps to avoid hash-collisions by telling whether an already occupied storage place has been occupied by the response to the stimulus just under consideration or whether it has been occupied by a response to another stimulus, meaning that the stimulus under generation is giving the same address just by chance. The second eight bits from the 16 bits not employed in generating the storage place address are then taken into account for generating some new address, hopefully not occupied by reactions to another stimulus.

For building the pseudo-random numbers out of the 32bit subelements three procedures mix1, mix2, mix3 have been written, allowing one to process three row vectors and/or twelve input elements s_i of $\underline{s}$, which proved to be sufficient up to now. If more than one row vector exists (more than four elements in $\underline{s}$) then the pseudo-random numbers generated by mixi are connected by an EXOR procedure, starting with mix1, mix2, and combining the result with the mix3 result, if more than eight inputs have to be considered. In EXOR the two numbers taken into account are added with the difference to a real addition, that the overflow of a "1" to the next higher bit-position, which appears if in a certain bit-position both numbers have a "1", is suppressed.

The mix-procedures themselves are a combination of the bit-manipulating functions SWAB, ROTL, and EXOR supplied by the software of the DEC-computers used in our laboratory. SWAB is exchanging the two bytes (8bit parts) in each of the 16bit halves. EXOR has been explained already. ROTL is a kind of rotation: all bits of a word and/or a 16bit half-word are shifted to the left by one bit, the now superfluous first bit being used as the now missing last bit.

As has been indicated already, the mix-procedures are applied only to the 16bit half-words of the 32bit subelements of the row-vectors. With a further adding of pseudo-random numbers and multiplications/divisions mod r by big numbers - mod r meaning neglecting the division rest of a division by the overall number of r storage places available and multiplications/divisions by big numbers being known to be a good tool for hash-coding - one gets finally fairly different addresses for the 16bit storage place addresses and for the 16bit identifiers plus address change pointers together with the required fairly uniform distribution of selected storage places about the physical memory provided. Actually the randomness achieved by this procedure is so good, that a "forced separation" for the r* places used to store the response - as it has been described in /10/ - was found to be not necessary any longer.

The main problem in building hash-coding procedures is to find the right compromise between reaching good randomness in storage place distribution and fast number manipulation, especially if one is interested in applying the associative memory for on-line control. The compromise is at least partly dependent on machine considerations and needs some practical experimentation, since the amount of randomness to be reached by a certain manipulation is difficult to predict.

The overall strategy is now as follows: the outputs to be stored are restricted to be 16bit-words. The storage places, providing 32bits, contain the respective output (response) element plus an 8bit identifier, connecting the output element with the input stimuli, and a counter, which shows how often a certain storage place has been actuated by a correct stimulus. One has to keep in mind in this respect that similar stimuli are using partly the same storage places leading to partly the same subelements in the row vectors, which are the basis for storage place calculation.

To discuss further details we have to make a distinction between the training and the recall mode of the memory. Both cases start with a calculation of the subelements of the row vectors from the stimulus content and with a further calculation of the storage place address and the identifier plus address change values from the subelement value in view of eventual hash-collisions.

In the training mode, it is checked first, whether the storage place obtained is occupied already, e.g. by looking at the counter. If it is free, the wanted response element will be written onto this place together with the identifier and the counter will be set to one. If the storage place is already occupied, the identifier calculated from the stimulus and the identifier stored will be compared. If they are equal, fine, the counter will be tuned up by one and the response element will be modified according to the training requirement. If the identifiers are different, a hash-collision has to be avoided. The calculated address will be changed by adding to it the also calculated 8bit address change filled up e.g. by zeros in the front part and for the new address the counter and identifier check is made again. If the new address is occupied and an identified difference is found to exist again, the adding up of the address change to the actual address is repeated, until either a non-occupied place is reached or the correct identifier or a certain limit of address changes is exceeded. In the last case some storage place content is superscribed, normally the last one or - as in our program - the first one in the row of addresses tested. The other cases are handled as already described for the case of no hash-collision.

In the recall or "association" mode of the memory, a counter value of zero means, that the respective stimulus/situation subelement has not been trained up to the moment. With a counter value unequal to zero, the identifier is the decisive element again: For an equal identifier value (stored and calculated), the response element stored will be activated, for an unequal value the search is continued by going to the next address through adding the address change value, until either an empty storage place is met, meaning no knowledge is available, or the right identifier, meaning recallable information is present, or the limit is exceeded, meaning again that no knowledge can be recalled.

II.2.4. The third mapping: Construction of the memory responses/outputs

As has been stated earlier, the response value to a certain stimulus is constructed from the contents of r^* memory cells called up by the stimulus values themselves. As an indication of how this is performed only a simple summation of the weights in the storage places has been mentioned up to now. However, one has to deal actually with two tasks. The first task is connected with the training mode and comprises the procedure to build up and/or change already present weights in the activated storage places. The second task is connected with the recall mode and deals with the question of how to combine the values from the activated storage places to obtain an appropriate response. We shall discuss the second question first, since it may seem at first glance, that there exists no problem at all. However, one has to take into account that a stimulus in the recall mode has not necessarily been encountered in the training mode. If there has been a lot of training in the area, where the recall stimulus is applied, all r^* storage cells being activated by the stimulus will have a certain content from training stimuli. By using them for a response calculation an interpolation in a certain sense will be achieved. However, if the training is not dense in the area considered, it may happen also that only some of the r^* memory cells will have been filled up by information from training stimuli. This is a very doubtful basis for estimating the correct response.

This leads one to the first question of how many of the r^* activated storage places have to be previously trained, so that the stimulus can be judged in the recall mode to be defined accurately enough to be usable for further calculations. Experience has shown, that one should require 60-80% of these r^* elements to be trained to call the stimulus response as a whole reliable. We will shortname the percentage of trained storage places η, η_{min} being by that the minimum percentage required to consider a certain stimulus situation as sufficiently trained.

However, even after this decision some further decision has to be made how to handle the untrained places among the r^* places activated. The correct procedure is to use just the weights in the $\tilde{r}^*$ trained storage places and keep the untrained $r^* - \tilde{r}^*$ storage places out of the calculation. That means, one has to check for each place at first, whether it contains a trained value or not. That is an easy task with the procedure discussed in the last section, in which the number of times that a certain storage place is trained with a meaningful value is counted. However, in the case of a low danger of hash-collisions - that means for a low percentage ($\leq 60\%$) of trained storage cells from the available memory - one can accelerate the storage procedure by dropping the calculation of identifiers and random address changes for places already occupied, and characterize the training status by using an additional memory of the same size and with the same activation mode as the memory used for information storage. This so-called "training indicator" is set to zero at all places in the beginning and just a "one" is trained into a storage place, when it is activated by the stimulus. In the recall mode, one gets by summing the content of the r^* activated cells a value $\tilde{r}^*$, indicating the number of trained places and from this the knowledge, whether the respective stimulus response can be considered as reliable or not. If, furthermore, the memory containing the information is starting with zero values in all storage places before training takes place, a summation of all weights in the r^* activated places will lead to the same result as summing the weights from the $\tilde{r}^*$ trained places, so that a check as to whether the places to be considered are trained or not trained can be avoided.

However, the training indicator allows such a check as well, since untrained places contain a zero and trained places a one. If the information memory is filled by random values in the beginning, the use of all r^* places may lead to wrong outputs, so the restriction to use trained places may be necessary in this case.

The training indicator scheme was implemented in some early versions of the learning control loop and proved to work very sucessfully.

Finally it should be remarked, that one can drop further the training indicator also, if the information accumulating memory has been trained densely enough in advance with meaningful storage place contents and on-line training is used only for improvements of this pre-trained information, since in this case one can always base the output construction on r^* memory places independently whether or not they have been addressed by on-line training.

In the following we will restrict ourselves to the case, where a check is made, whether or not the considered memory places are already trained. If we go back to fig. 11, we remember that to each of the rectangles shown in the lower part some storage place is assigned, in which a certain weight w_{ij} may be stored for the output p_i or at a restriction onto one output p a weight w_j. With the definition

$$
(5a) \qquad \alpha_j = \begin{cases} 1 \text{ if } w_j \text{ is a trained value} \\ 0 \text{ if this is not the case} \end{cases}
$$

we get for a considered stimulus $\underline{s}$:

$$
(5b) \qquad \sum_{j=1}^{r^*} \alpha_j = \tilde{r}^*
$$

Two different procedures for construction of the response to stimulus $\underline{s}$ have been considered and successfully applied:

(6a) $$p = \frac{1}{r^*} \bullet \sum_{j=1}^{r^*} \alpha_j w_j \quad {}^2$$

(6b) $$p = \frac{1}{r^*} \bullet \sum_{j=1}^{r^*} \alpha_j w_j$$

To clarify the differences let us assume that the result of weight selection is $w_j = p^{desired}$ for all j and that just one stimulus has been trained as shown in fig. 11 by the shaded regions in the coarse grids. A recall with the same stimulus as trained will give for both formulas the correct answer $p^{desired}$, since all r^* elements of the coarse grid activated by the stimulus are trained. However, for a stimulus at a certain distance from the training stimulus, which activates, for example, only two of the trained rectangles from the coarse grid, one obtains from (6a) $p = 2/r^* \bullet p^{desired}$ but from (6b) $p = 2/2 \bullet p^{desired} = p^{desired}$. As shown in fig. 12 this means, that in the first case the basic elements for forming the stimulus-response surface are quantized cones and in the second case plateaus.

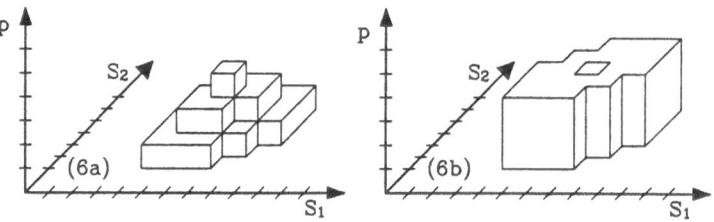

<u>Fig. 12:</u> Results of response (output) - construction from (6a) and (6b) - $r^* = 3$, one trained stimulus $\underline{s}' = (s_1, s_2)$, one output p.

Both versions can be combined with the same alternatives for weight updating. In this connection we shall start with the assumption, that either noise is small in the measured data or that we have not implemented the counters in the hash-coding procedure, which is a necessary condition for the filtering procedure to be discussed later on.

---

[2]Actually the index numbers j are defined by the respective hypercubes. Therefore one should correctly specifiy the set of indices of activated storage cells for a certain stimulus $\hat{\underline{s}}$ by say $A(\hat{\underline{s}})$ and write here and later on $\sum_{j \in A(\hat{\underline{s}})}$.

Two main lines of weight updating have been followed. In the simplest approach the untrained places are set to the desired output value and the trained places are corrected equally in the sense that the desired output is reached by summing all r^* values of the activated memory cells.

(7a) $\qquad w_j^{new} = p^{desired}$ $\qquad\qquad\qquad\qquad$ for untrained places

(7b) $\qquad w_j^{new} = w_j^{old} + [\, p^{desired} - \dfrac{1}{\tilde{r}^*} \sum_1^{r^*} \alpha_j w_j^{old}\,]$ $\qquad\qquad$ for trained places

Since now all r^* places are trained, one obtains as output:

(7c) $\qquad p = \dfrac{1}{\tilde{r}^*} \sum_{j=1}^{r^*} w_j^{new} = \dfrac{1}{r^*} \cdot [\, r^* p^{desired} + \sum_1^{r^*} \alpha_j w_j^{old} - \sum_1^{r^*} \alpha_j w_j^{old}\,] = p^{desired}$

As long as there is no interaction between the fields of local generalization of different stimuli, this is perfectly satisfactory. However, overlapping stimulus influence regions mean that the weights implemented by the stimulus trained first are changed partly by a stimulus trained later on, meaning that now the first stimulus does not give the exact output any longer if called up again. Therefore it seems more appropriate to keep the changes to weights trained already as small as possible, which leads to the most smooth interpolation and can be achieved by distributing the difference between actual and desired output equally over all r^* activated cells. The respective algorithmus can be written:

(8a) $\qquad w_j^{new} = p^{desired} + \dfrac{1}{r^*} [\, \tilde{r}^* p^{desired} - \sum_1^{r^*} \alpha_j w_j^{old}\,]$ $\qquad\qquad$ for untrained places

(8b) $\qquad w_j^{new} = w_j^{old} + \dfrac{1}{r^*} [\, \tilde{r}^* p^{desired} - \sum_1^{r^*} \alpha_j w_j^{old}\,]$ $\qquad\qquad$ for trained places

giving

(8c) $\qquad p = \dfrac{1}{r^*} \sum_1^{r^*} w_j^{new} = \dfrac{1}{r^*} [\, (r^* - \tilde{r}^*) p^{desired} + \sum_1^{r^*} \alpha_j w_j^{old} + \tilde{r}^* p^{desired} - \sum_1^{r^*} \alpha_j w_j^{old}\,]$

$\qquad\qquad\qquad = p^{desired}$

If counters, which indicate how often a certain storage place has been activated by stimuli, are implemented a meaningful weight updating is:

$$(9a) \qquad w_j^{new} = w_j^{old} + \frac{1}{k_j+1} (p^{desired} - p^{old})$$

k_j being the value of the counter, which is incremented by 1 each time the storage place is activated. Defining the mean weighting factor as:

$$(9b) \qquad \frac{1}{k} = \frac{1}{r} \cdot \sum_{1}^{r} \frac{1}{k_j+1}$$

one obtains for the response:

$$(9c) \qquad p = \frac{1}{r} \sum_{1}^{r} w_j^{new} = p^{old} + \frac{1}{k} (p^{desired} - p^{old})$$

which means that each training step changes the output for the respective training stimulus by a fraction of the difference between the desired and the actual output, the fraction being determined by the mean weighting factor $\frac{1}{k}$. Since $\frac{1}{k}$ is declining with the number of training steps, the output value gets more and more resistant to changes with a growing number of training repetitions.

In /11/ it is shown, that if the response value to be trained is $\tilde{p} = p^{desired} + n(t)$ with $n(t)$ noise with the mean value zero, the response value tends to $p^{desired}$ as the number of training cycles goes to infinity. In this sense , (9a) leads to a filtering property. However, one has to be careful about the significance of such a filter algorithmus. If one uses the locally generalizing associative memory to learn a non-linear characteristic of a technical process by training the memory to the output according to some specified, repeated input sets, this procedure of suppressing noise effects is very effective. If one considers, however, learning control loops as presented in a first, rough version in section I.4, the plant situation (stimuli) for the predictive model comprises in general input and output values of earlier points from the time history, which means noise from the plant output measurements is an inevitable element of the predictive model inputs and the consequential effects can be levelled out only by the generalization property of the memory.

A further point to be mentioned is, that one cannot distinguish from the measurement at a certain moment, whether an unexpected process answer is due to measurement noise or due to some change of the process itself. So the low weighting of differences between the old response and the measured ("desired") response as used in (9a) makes the adaptation of process models to process changes very difficult. Some possibility to counter-act the growing stiffness of stored information without loosing the filtering possibility fully is, to reduce the contents of the counters after a certain time by some value up to anulling by this very old information totally (that is introducing a "forgetting factor").

A general rule as to which of the weight updating rules (7), (8), (9) is the best one cannot be given. The respective decision has to be made in connection with the envisaged application.

II.2.5. Performance

AMS has been used on the one hand for research work on the properties of such a locally generalizing associative memory system but also on the other hand as a basic element of learning control loops. This has led over the years to different implementations of varying complexity on different computers, making detailed performance statements fairly difficult to write down. However, some rough figures will be given, which may be considered to be representative for the layout described above.

Starting with the necessary memory space, one has to distinguish between the portion necessary for data storage and the portion necessary for AMS organisation and data handling.

The amount of memory places necessary for data storage depends naturally on the stimulus-response connection just considered. By distributing the response values on r^* places instead of storing it in one place one gets, however, a considerable storage place reduction, which can be characterized in the following way: Let us assume that we have n stimuli and that the quantizations ϵ_i for all n stimuli have been fixed in such a way, that we get the same amount R of basic intervals for the respective ranges s_{imax}-s_{imin}. Then the original grid due to the quantization would require

$$(10a) \qquad z_b = R^n$$

storage places, each hypercube of the grid requiring just one memory place for storing the response value. Using r^* coarser grids with interval length $r^* \cdot \epsilon_i$ one needs, however, only

$$(10b) \qquad z_c \approx r^* \cdot \left[\left(\substack{\text{next} \\ \text{integer}} \frac{R}{r^*}\right)+1\right]^n$$

weight storage places. In (10b) next integer means the first multiple of r^* being bigger than R and the +1 in the bracket takes into account, that the shifted intervals of the r^* coarser grids for the n stimuli are extending over the limits s_{imax} and/or s_{imin}, so that in general just one interval more is required, than would be necessary to include $s_{imax}-s_{imin}$ without the fixed shifting (compare fig. 11).

Now, from (10a), (10b) as the relative reduction percentage of required storage places for distributed storage against normal storage follows:

$$(10c) \qquad 100 \cdot \frac{z_c}{z_b} = \frac{r^*[(^{next}_{integ.}\frac{R}{r^*}) + 1]^n}{R^n} = z \ \%$$

Table 1 shows some values calculated by (10c) and demonstrates that in real applications dramatic reductions may be achieved.

		n = 2	n = 4	n = 8
$r^* = 8$	R = 100	16 %	0,3 %	$0,1 \cdot 10^{-3}\%$
	R = 1000	13 %	0,2 %	$0,05 \cdot 10^{-3}\%$
$r^* = 16$	R = 100	10 %	0,07 %	$0,3 \cdot 10^{-5}\%$
	R = 1000	7 %	0,03 %	$0,04 \cdot 10^{-5}\%$
$r^* = 32$	R = 100	8 %	0,02 %	$0,1 \cdot 10^{-6}\%$
	R = 1000	3 %	0,004 %	$0,004 \cdot 10^{-6}\%$

Table 1 - Reduction of storage requirements by distributed information storage (1 response/output)

Since the reduction of necessary memory space by distributed storage of the response values can be calculated fairly exactly, the further reduction by hash-coding is difficult to estimate. It depends on the density of actual stimuli in the estimated stimuli ranges. For equally just x% active stimuli ranges from the estimated safe stimuli ranges one would be led to

$$(11) \qquad 100(\frac{x}{100})^n = z_n \ \%$$

from the necessary storage place for distributed information storage, if the x% are compact parts of the stimuli ranges and if the size of the coarse grids is small with respect to these compact parts. Otherwise the two amounts of reduction cannot be added up simply. However, it should be remembered, that hash-coding was not used in the first place on account of possible memory size reduction reasons, but mostly for easy generation of storage place addresses and that it has furthermore the helpful property of lessening the negative effects of damage to part of the memory.

Fig. 13 exemplifies the comments made up to now. Fig. 13a shows the surface to be stored in the associative memory. Using a basic grid of the input variables s_1, s_2 with resolution one and the ranges $0 \leq s_1 \leq 256$, $0 \leq s_2 \leq 256$ one has $257 \cdot 257 = 66049$ grid elements. After a single training cycle with a local generalization $r^* = 16$ one gets with 289 equally distributed ($= 0,4\%$ of the possible) grid elements the surface shown in fig. 13b, which can be called a good approximation, the mean square error being just $2,1\%$. An erasion of 20% from the supplied 8192 memory places ($\approx 12,5\%$ of the 66049 grid elements) leads to fig. 13c with an in general still tolerable mean square error of $6,5\%$.

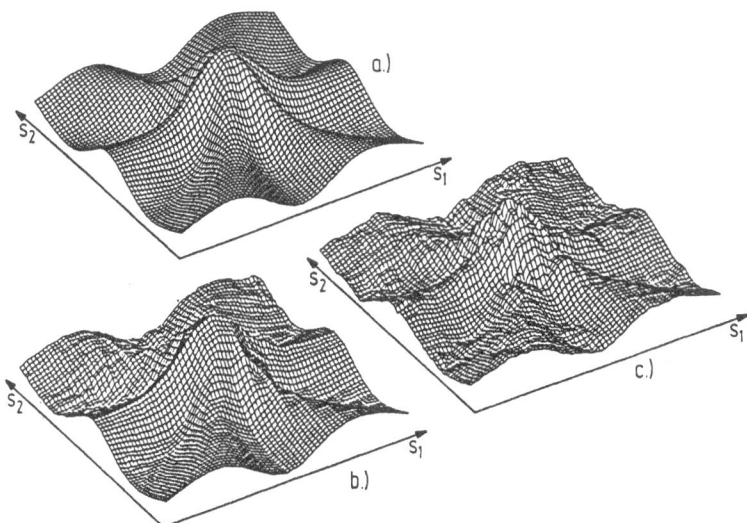

Fig. 13: Storage performance of AMS. a.) original surface; b.) reproduction from AMS with 8192 storage places, a local generalization of $r^* = 16$ and 289 out of 66049 original grid elements trained with the correct response value ($0,4\%$); mean square error $2,1\%$; c.) memory response after erasion of the stored weights from 20% of the 8192 memory places - mean square error from this partly destroyed memory: $6,5\%$.

The program for handling the supplied storage places in the AMS-fashion needs an amount of 12000 16bit words. The time for one association - training or recall of one stimulus/response connection - lies in the order of milliseconds. Using a PDP 11/73, which needs roughly 5 μsec for additions and 11 μsec for multiplications in fixed point arithmetics as used here (21 μsec and/or 34 μsec in the non used floating point arithmetic) one gets the association times of table 2 with the shown dependence from the numer n of stimuli and from the numer r^* as indicator of the amount of generalization.

	$r^* = 8$	$r^* = 16$	$r^* = 32$
$n = 3$	2 msec	3 msec	5 msec
$n = 6$	3 msec	5 msec	6 msec

Table 2 - Basic time spans for association between n stimuli and one response for different n and different generalization degrees r^*

One has to take into account, however, that the association times depend also on the degree, to which the memory is used. The numbers given in table 2 apply for a memory utilization of less than 40 %. With higher degrees of memory utilization, hash collisions are more and more frequent, leading to some search effort for free storage places. It is estimated that for an utilization degree of 80 % and an allowed search for a free storage place of up to 10 steps the time spans cited in table 2 may have to be doubled.

II.3. Results with some test functions

II.3.1. Test functions and quality assessment criteria

To evaluate the quality of locally generalizing associative memories regarding stimulus-response connection representation and interpolation accuracy one has to establish adequate tools. With this aim J. Militzer developed in 1985/86 (assisted by Z. Matios) the following strategy:

At first some (mostly two-dimensional) test functions with different characteristics were defined. Then a number ζ of equally spaced test points $\tilde{s}_k$ from the basic definition area of the test functions were selected and after the training of a certain number (30, 100, 300 ...) of randomly from an equal distribution chosen training points $\hat{s}_l$, the achieved memory responses $\hat{p}(\tilde{s}_i)$ were compared with the known exact function values $p(\tilde{s}_i)$ at the test points [3] by different criteria. Fig. 14 illustrates this procedure.

---

[3] It should be remembered that the hypercubes from the stimulus space are actually represented in all calculations by a certain one of their corners as discussed in section II.2.2.

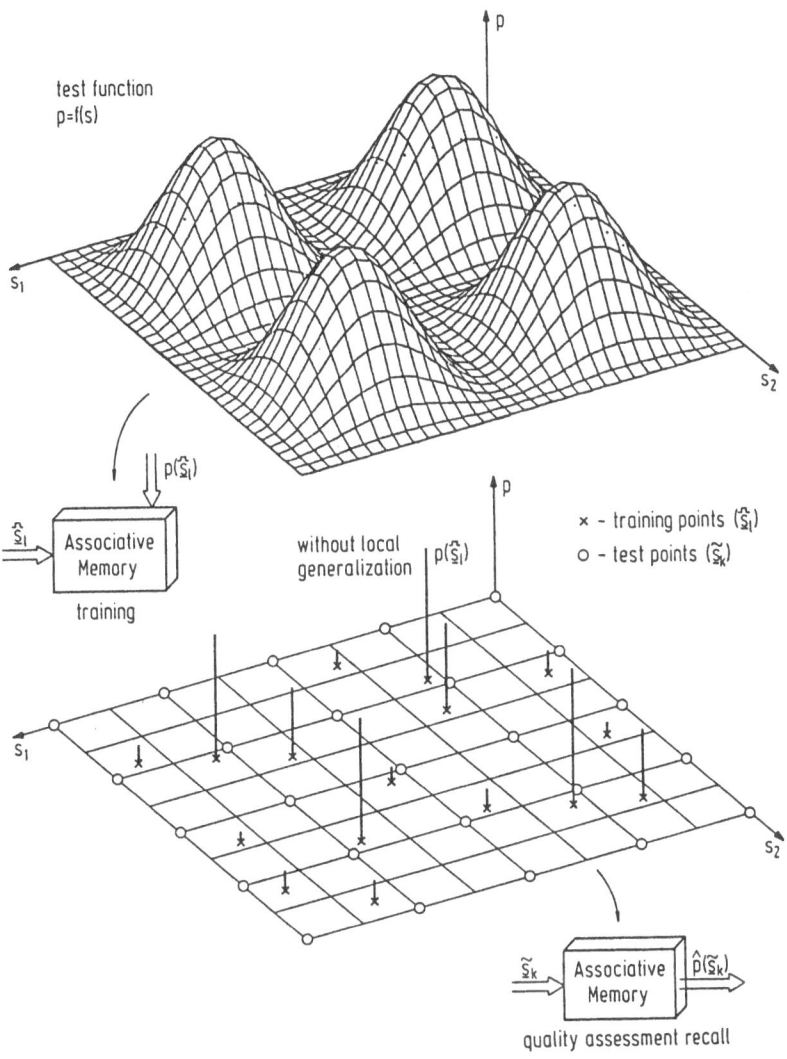

Fig. 14: Strategy for stimulus-response connection representation and interpolation accuracy quality assessment: A mathematically defined test function $p = f(s)$ is stored through a certain number of randomly chosen training points $\hat{s}_l$ by using the exact function value $p(\hat{s}_l)$ as training value. The locally generalized - here not drawn - associative answer $\hat{p}$ is recalled in equally spaced test points $\tilde{s}_k$ and compared with the exact values $p(\tilde{s}_k)$.

In the criteria used a distinction has been made between the ζ test points in general and test points judged as representing a trained value being characterized by η_{min} (compare section II.2.4). The respective number χ of trained test points can be calculated by:

$$(12) \qquad \chi(\eta_{min}) = \sum_{i=1}^{\zeta} \tilde{a}_i \text{ with } \tilde{a}_i = \begin{cases} 0 \text{ for } \eta_i < \eta_{min} \\ 1 \text{ for } \eta_i \geq \eta_{min} \end{cases}$$

Using (12) the selected quality criteria can be written:

Mean absolute error with respect to the asolute response value

$$(13) \qquad E_{bm}(\eta_{min}) = \frac{1}{\chi} \sum_{i=1}^{\zeta} \tilde{a}_i |\hat{p}(\tilde{s}_i) - p(\tilde{s}_i)| \left[\frac{1}{\zeta} \sum_{i=1}^{\zeta} |\hat{p}(\tilde{s}_i)| \right]^{-1}$$

Mean square error with respect to the mean square response value:

$$(14) \qquad E_{eff}(\eta_{min}) = \left[\frac{1}{\chi} \sum_{i=1}^{\zeta} \tilde{a}_i (\hat{p}(\tilde{s}_i) - p(\tilde{s}_i))^2 \right]^{1/2} \left[\frac{1}{\zeta} \sum_{i=1}^{\zeta} (\hat{p}(\tilde{s}_i))^2 \right]^{-1/2}$$

Maximal absolute error:

$$(15) \qquad E_{max}(\eta_{min}) = \frac{max}{i=1,2..\zeta} \left[\tilde{a}_i |\hat{p}(\tilde{s}_i) - p(\tilde{s}_i)| \right] \cdot \left[\frac{max}{i=1,2..\zeta} |\hat{P}_i(\tilde{s}_i)| \right]^{-1}$$

Actually nine different test functions - partly taken from a catalogue given by H. P. Schwefel in /12/ - were taken into account for a systematic research on memory qualities. We shall restrict the demonstration of results achieved to the following three test functions defined in all cases over the stimulus area $0 \leq s_i \leq 4096$ with $i=1,2$; $x_i = s_i/4096$):

$$(16) \qquad \text{FSIN:} \qquad p(s_1,s_2) = \sin^2(2\pi x_1) \cdot \sin^2(2\pi x_2)$$

$$(17) \qquad \text{FCOS:} \qquad p(s_1,s_2) = \frac{\cos\{2\pi[(2x_1-1)^2 + (2x_2-1)^2]\}}{e^{1/4} \cdot [(2x_1-1)^2 + (2x_2-1)^2]}$$

$$(18) \qquad \text{CORNER:} \qquad p(s_1,s_2) = \begin{cases} 0,75x_1 \text{ for } x_1 < 0,6; x_2 > 0,5 \\ 1 \qquad \text{otherwise} \end{cases}$$

Sketches of the three function can be found in figures 15a, 16a and 16c.

With respect to FSIN has FCOS the property of a stronger curvature, the function CORNER contains on the other hand real jumps.

II.3.2. Effects of various parameter variations

The effects, which we are going to discuss, are the results of a variation of the degree of local generalisation r^*, the results of a variation of the maximum number of allowed steps in the search for free memory places in case of hash collisions σ and the results of a variation in the response construction as described by equations (6a), (6b) - see also fig. 12 -.

We shall start with some pictures regarding the effects of r^* variation and turn then to the quality measures defined in the last section.

The test functions are considered as being given by a calculation of $p(\underline{s})$ at the points $s_i = 0,1,2...4096$, that means at 4097x4097 = 16 785 409 points. In the following pictures, however, a basic resolution of $\epsilon_1 = \epsilon_2 = 16$ points was used for storage in the AMS-system, which leads for $r^* = 8$ to $\epsilon \cdot r^* = 128$ and for $r^* = 16$ to $\epsilon \cdot r^* = 256$ as the size of the 8 and/or 16 element coarse grids shifted by ϵ against each other.

Fig. 15 demonstrates - using the response construction of equation (6a) - the effects of a r^*-variation in dependence of the number of trained points. With $\epsilon = 16$ the total number of basic grid-elements and/or grid-element representing corner points is roughly 65 500. Fig. 15a gives a graphic representation of $p(\underline{s}) = $ FSIN, fig. 15b $\hat{p}(\hat{\underline{s}}_i)$ for 1000 training points with $r^* = 8$ and fig. 15d $\hat{p}(\hat{\underline{s}}_i)$ for 10 000 training points and the same r^*. One sees, that an $r^* = 8$ is not able to represent FSIN by 1000 training steps, equal 1,5 % of the overall 65 500 points. However, fig. 15c shows, that this is the case with $r^* = 16$. Fig 16 demonstrates that a good representation of FCOS - as shown in fig. 16a - can be reached as well with $r^* = 16$ and 1000 training steps, and that this is also the case for the CORNER given in fig. 16c - as exemplified in fig. 16d.

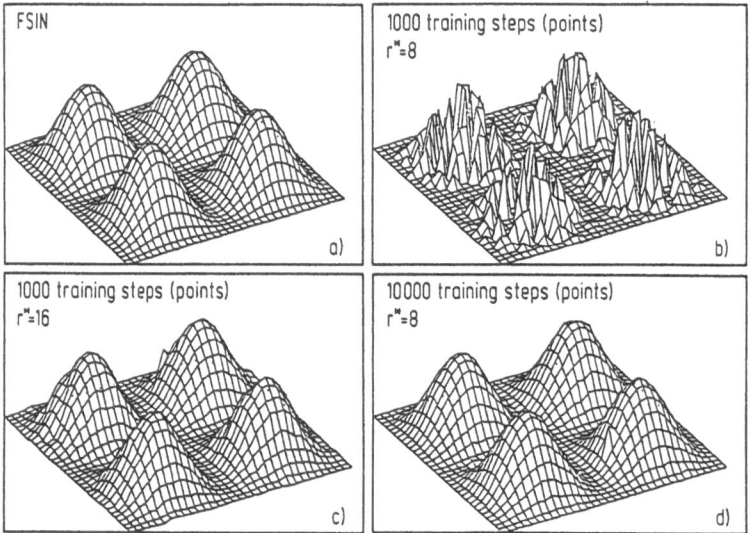

Fig. 15: Effect of variation of r^*. For $1000 \approx 1.5\ \%$ trained points from the 65 500 points representing FSIN, $r^* = 8$ does not give a good description of the original function - compare b) -, however, $r^* = 16$ does - compare c).

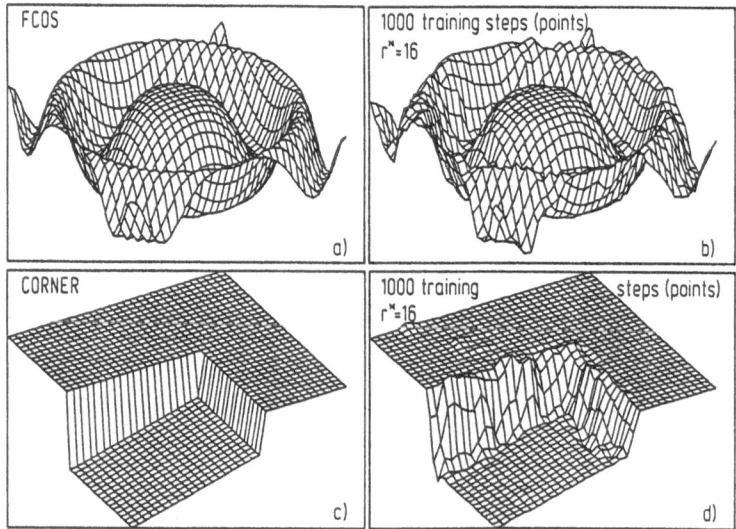

Fig. 16: Verification of results from fig. 15 regarding the sufficient representation of original functions through 1000 training steps and $r^* = 16$ for the functions FCOS and CORNER.

Naturally, the quality of function representation depends not only on r^* but also on the whole number r of storage places supplied and the allowed number of searches σ for a free place before the actual data are written into an already occupied place in case of hash collisions. Furthermore the value of η_{min}, deciding which $\underline{s}_i$ are considered to be trained, plays some role.

In the following tables these effects will be discussed. The test points $\tilde{s}_k$, for which the criteria (16) - (18) are calculated, are spaced with a difference of 32 from the 4097 points in each stimulus direction, so that one gets 129x129 = 16 641 test values to be taken into account.

Table 3 deals with the question of available storage places r for FSIN and an equal $r^* = 16$. $(r^* \cdot \epsilon = 256)$. The response construction is here (and hereafter unless stated otherwise) as explained in equation (6a). This table and the following tables are structured by the number of training steps (training points) and the value of η_{min} used. Besides the values of the criteria (16) - (18) there is always additionally indicated the percentage of test points considered untrained due to the used value of η_{min}.

available storage places	number of training points	$\eta_{min} \geq$	E_{bm}	E_{eff}	E_{max}	untrained test points
49,3 KByte	1 000	76 % 82 % 94 %	5,57 % 5,50 % 5,50 %	5,22 % 5,10 % 5,07 %	11,00 % 11,00 % 9,00 %	6,2 % 10,0 % 19,5 %
	10 000	82 % 94 %	3,11 % 3,11 %	2,77 % 2,77 %	4,50 % 4,50 %	0,0 % 1,5 %
14,7 Kbyte	1 000	76 % 82 % 94 %	5,62 % 5,55 % 5,55 %	5,26 % 5,15 % 5,13 %	11,30 % 11,30 % 9,00 %	6,2 % 10,1 % 19,6 %
	10 000	82 % 94 %	3,15 % 3,15 %	2,80 % 2,80 %	5,80 % 5,80 %	0,0 % 1,9 %
8,75 KByte	1 000	76 % 82 % 94 %	6,22 % 6,19 % 5,88 %	5,78 % 5,73 % 5,26 %	11,10 % 11,10 % 10,20 %	8,3 % 22,8 % 47,9 %
	10 000	82 % 94 %	3,69 % 3,60 %	3,27 % 3,15 %	6,30 % 4,80 %	4,6 % 46,3 %

Table 3 - Results of a change on overall available storage places - FSIN, $\sigma = 10$, $r^* = 16$, $\epsilon = 16$

Looking at table 3 one sees, that the number of test points classified as untrained is, on the one hand dependent on η_{min} and the number of used points in the training phase as expected, but that additionally this number rises if the available storage volume is becoming fairly small. This is due to the fact that in a small memory hash collisions, even with the search depth for free memory places of $\sigma = 10$ used here, are frequent and that by such a hash collision some already correctly trained information is destroyed due to the local generalization - not only in one place but in r^* places. So instead of getting a clearly smaller amount of untrained test points with a growing number of training points, as it is the case for the larger memories, one gets for $\eta_{min} \geq 94\ \%$ for the 8,75 KByte memory a stationary percentage of untrained test points. That this has no effect on the quality criteria is due to the fact that these criteria consider the errors in the trained test points only. As an overall conclusion one can come to the opinion from table 3 that for the considered 65 500 possible points and $\sigma = 10$ a memory size of roughly 15 KByte is necessary and sufficient.

Table 4 shows now the result of training FSIN under the same conditions as in table 3, but for $\sigma = 3$.

available storage places	number of training points	η_{min} $\geq$	E_{bm}	E_{eff}	E_{max}	untrained test points
49,3 KByte	1 000	76 %	5,57 %	5,22 %	11,00 %	6,2 %
		82 %	5,50 %	5,11 %	11,00 %	10,0 %
		94 %	5,50 %	5,07 %	9,00 %	19,8 %
	10 000	82 %	3,11 %	2,77 %	4,50 %	0,1 %
		94 %	3,11 %	2,77 %	4,50 %	1,8 %
14,7 Kbyte	1 000	76 %	5,65 %	5,29 %	11,20 %	7,0 %
		82 %	5,58 %	5,18 %	11,20 %	11,6 %
		94 %	5,45 %	5,04 %	9,10 %	36,0 %
	10 000	82 %	3,20 %	2,85 %	4,60 %	0,6 %
		94 %	3,14 %	2,78 %	4,50 %	17,9 %

Table 4 - Training of FSIN under the same conditions as in table 3 with the exception of here $\sigma = 3$ instead of $\sigma = 10$ in table 3.

The main difference to table 3 lies in the amount of untrained points for the 14,7 KByte memory: One sees that for economic handling of necessary storage amount an identifier with the connected possibility for search after free storage places is in fact important and that even a relatively high depth of search, like 10 maybe required. To go much further does not make sense for two reasons: first it may delay the association time too much. Secondly in general the available different identifiers are limited (in our case up to 255 different identifiers), so that with high σ-values the probability grows, that one reaches the same identifier, however, characterizing different data.

Let us now turn back to the question of adequate local generalization, as discussed visually already by the pictures in fig. 15 and 16. However, it has to be remarked first that there are two ways of changing the amount of local generalization: one may extend ϵ or one may extend r^*. Since both methods mean the same for a response construction according to (6b), the second measure means a differentiation into more weights for the same area and by this the ability of better adaptation to curvatures in case of a response construction by formular (6a) - see also fig. 12 -. Table 5 supports this statement. It shows results for $\epsilon=32$, $r^*=16$ giving $\epsilon r^*=512$ and for $\epsilon=16$, $r^*=16$; $\epsilon=32$, $r^*=8$ giving $\epsilon r^*=256$ as well as for $\epsilon=8$, $r^*=16$; $\epsilon=16$, $r^*=8$ giving $\epsilon r^*=128$.

available storage places		number of training points	η_{min} $\geq$	E_{bm}	E_{eff}	E_{max}	untrained test points
$\epsilon r^*=128$	$\epsilon=8$	1 000	76 %	4,11 %	3,91 %	6,90 %	65,3 %
			88 %	3,54 %	3,32 %	6,10 %	78,0 %
	$r^*=16$	10 000	88 %	1,95 %	1,88 %	6,60 %	2,2 %
	$\epsilon=16$	1 000	76 %	4,52 %	4,26 %	7,80 %	68,3 %
			88 %	4,00 %	3,71 %	6,30 %	81,1 %
	$r^*=8$	10 000	88 %	3,01 %	2,79 %	7,50 %	3,3 %
$\epsilon r^*=256$	$\epsilon=16$	1 000	76 %	5,62 %	5,26 %	11,30 %	6,2 %
			88 %	5,54 %	5,12 %	9,00 %	15,3 %
	$r^*=16$	10 000	88 %	3,15 %	2,80 %	5,80 %	0,2 %
	$\epsilon=32$	1 000	76 %	7,14 %	6,45 %	14,10 %	8,7 %
			88 %	7,14 %	6,45 %	14,10 %	18,2 %
	$r^*=8$	10 000	88 %	5,5 %	4,92 %	7,80 %	1,3 %
$\epsilon r^*=512$	$\epsilon=32$	1 000	76 %	8,31 %	7,17 %	13,10 %	0,1 %
			88 %	8,31 %	7.17 %	13,10 %	0,7 %
	$r^*=16$	10 000	88 %	7,76 %	6,63 %	10,20 %	0,0 %

Table 5 - FSIN results for different degrees of each generalization achieved by different basic quantization ϵ and/or different values of r^* ; $\sigma=10$.

One sees, that for the same value of ϵr^* one gets better quality values for smaller ϵ and bigger r^*. Growing ϵr^* gives for the same number of training points and same training indicator η_{min}-values smaller amounts of untrained test points, paid for by a slightly less exact imitation of the original surface.

Table 6 gives some further details on the quality of the surface descriptions shown in fig 16 with 1000 training steps. One sees, that the value of η_{min}, which determines the number of points considered to be untrained and being therefore not used in the quality criteria evaluation, is of very low influence regarding mean errors as well as maximum errors. A much greater influence has the degree of curvature and/or gradient steepness. Especially the maximum error shows the modelling problems for strong gradient changes, a result to be expected for the principle of local generalization, which implicitly assumes a certain smoothness of surfaces.

	η_{min}	E_{bm}	E_{eff}	E_{max}	untrained test points
FSIN	51 %	5,76 %	5,47 %	12,70 %	0,9 %
	69 %	5,69 %	5,35 %	11,30 %	4,0 %
FCOS	51 %	9,53 %	12,60 %	60,90 %	0,9 %
	69 %	9,11 %	11,82 %	50,80 %	4,0 %
CORNER	51 %	2,58 %	8,23 %	81,50 %	0,9 %
	69 %	2,48 %	7,89 %	81,50 %	4,0 %

Table 6 - Comparison of quality criteria value for functions of different quality, $\epsilon=16$, $r^*=16$, $\sigma=10$, memory size 14,7 KByte, 1000 training points

One could assume, that this situation may be improved by using the second method of response construction not considered until now, since jumps seem easier to be handled this way. However, this is not the case. Table 7 give the results of performing the same calculations necessary for constructing table 6, but by using formula (6b) instead formula (6a) for calculating the response values. One sees, that the results are even slightly worse than those of table 6. This is due to the fact, that the degree of generalization is everywhere the same, which means a point just at the edge, where the jump in the corner occurs, leads to an unwanted extension of the plateau. So possible advantages in forming plateaus are lost at edges, leading to high error accumulation in the performance criteria and through this to the results of table 7.

	η_{min}	E_{bm}	E_{eff}	E_{max}	untrained test points
FSIN	51 %	6,44 %	7,42 %	30,30 %	0,9 %
	69 %	6,38 %	7,10 %	22,60 %	4,0 %
FCOS	51 %	11,18 %	14,88 %	57,00 %	0,9 %
	69 %	10,24 %	13,18 %	50,00 %	4,0 %
CORNER	51 %	5,69 %	12,01 %	81,50 %	0,9 %
	69 %	4,62 %	10,08 %	81,50 %	4,0 %

Table 7 - Calculation of response values with formula (6b) instead with formula (6a) basic parameters as stated for table 6 .

II.3.3. Noise filtering

In section II.2.4. it was mentioned, that if counters are implemented together with hash-coding they can be used to reduce noise effects via the weight updating formula (9a). Now some numerical results to support this statement will be presented.

The same assumptions and/or parameters are employed as for computing table 6, so this table can be considered as giving reference results for stochastic noise with the mean value zero and a standard deviation of 0 %. Table 8 shows the respective values for a standard deviation of 20 %.

Function	filtering	number of training points	η_{min}	E_{bm}	E_{eff}	E_{max}	untrained test points
FSIN	without	1 000	69%	49,19%	40,47%	54,50%	4,0 %
			82%	49,23%	40,44%	54,50%	10,0 %
		10 000	82%	51,55%	42,65%	55,00%	0,0 %
	with	1 000	69%	24,70%	20,75%	27,80%	4,0 %
			82%	24,39%	20,45%	27,80%	10,0 %
		10 000	82%	9,32%	7,80%	11,50%	0,0 %
FCOS	without	1 000	69%	24,10%	26,78%	59,80%	4,0 %
			82%	23,77%	26,32%	53,70%	10,0 %
		10 000	82%	24,59%	27,43%	58,00%	0,0 %
	with	1 000	69%	18,94%	21,60%	59,60%	4,0 %
			82%	18,30%	20,64%	53,00%	10,0 %
		10 000	82%	13,39%	15,06%	50,30%	0,0 %
CORNER	without	1 000	69%	16,65%	19,04%	89,90%	4,0 %
			82%	16,67%	19,01%	89,90%	10,0 %
		10 000	82%	17,57%	20,14%	95,30%	0,0 %
	with	1 000	69%	8,54%	10,97%	91,30%	4,0 %
			82%	8,44%	10,86%	91,30%	10,0 %
		10 000	82%	4,01%	7,46%	56,50%	0,0 %

Table 8 - Results for response construction according to (6a), search depth σ=10, quantization ϵ=16, generalization r^* =16, memory size 14,7 KByte and stochastic noise with mean value zero and standard deviation of 20 %.

By analyzing table 8 one sees immediately, that with 1 000 training points one doesn't reach a good noise reduction through the filtering procedure: although, due to the overlapping of the local generalization fields, certainly most of the test point grid elements have been activated more than once so that some filtering effect can be recognized, this is still far from being satisfactory. However with 10 000 training points results can be achieved, which are not far off from the results, which one gets in case of no noise with 1 000 training points. This result seems to be sufficient at least when judged from the sketches in figures 15 and 16.

So noise filtering can be performed by the AMS with the right weight change procedure and using counters, however, it has to be paid for by an increased training effort.

<u>II.4. Theoretical considerations</u>

<u>II.4.1. Basic properties</u>

In principle locally generalizing associative memories generate nonlinear manifolds by linear superposition of basic functions, namely those sketched in fig. 12. As a theoretical background for such a procedure one should remember that any quadratic integrable function can be represented over a certain definition area by a complete set of linearly independent quadratic integrable functions defined over this area. This fact has been exploited especially for the solution of differential equations, which can be derived from a minimization problem. E.g. surfaces characterized by minimum energy are examples which can be named in this context. Actually the most efficient tool for calculating such surfaces is the finite element method and we shall therefore make a short comparison between this way of surface representation and the AMS approach in handling this task. To minimize the complexity of mathematics and pictures, we will restrict ourselves to problems of one dimension.

According to /13/ we can define a linear finite element $\varphi_k^h(s)$ over a coordinate s with the spacing h of basic points stretching from $0 \cdot h$ till $N \cdot h$ by fig 17: $\varphi_k^h(s)$ is given the value 1 at $s = k \cdot h$ and the value 0 at all other $s = i \cdot h$, $i \neq k$ with a linear connection from the value one to the values zero at $(k-1) \cdot h$ and $(k+1) \cdot h$, making $\varphi_k^h(s)$ continuous over the definition interval.

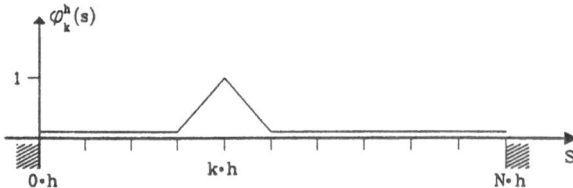

<u>Fig. 17:</u> Definition of linear continuous finite element $\varphi_k^h(s)$

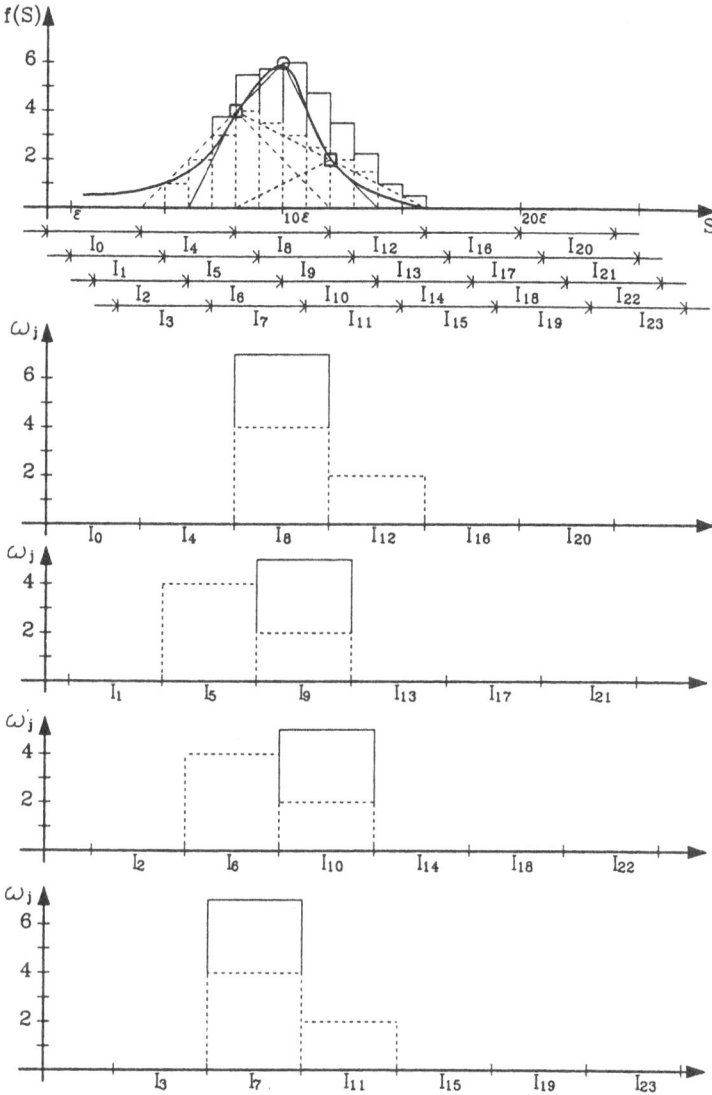

Fig. 18: Approximation of f(s) by finite elements and AMS (6a), (8a,b) with r*=4. Broken lines trained with points □ at 8ε, 12ε, full lines with later on added point at 10ε ⇒ h=2 for the finite elements and/or weight changes due to (8a,b) for AMS as indicated in the lower part of the figure.

By the superposition law:

$$(19) \qquad f(s) = \sum_{k=1}^{N} q_k \, \varphi_k^h(s) \equiv \sum_{k=1}^{N} f(k \cdot h) \, \varphi_k^h(s)$$

any function $f(s)$ will be approximated by a linear interpolation between the values at $s = k \cdot h$, since (19) gives the exact values $f(k \cdot h)$ at these points due to the construction of the $\varphi_k^h(s)$ and since furthermore the superposition of linear functions - as the $\varphi_k^h(s)$ are between the considered points $i \cdot h$ - gives again a linear function.

The distance h may be equal to the quantization value ϵ but it may be also some multiple of ϵ. Let us compare the approximation of some function $f(s)$ by the finite element method with linear elements and the AMS output using formula (6a), which gives according to fig. 12 and fig. 8 terraced triangles as basic functions. Results are shown in fig. 18. If we use at first $f(s_1)=f(8\epsilon)=4$ and $f(s_2)=f(12\epsilon)=2$ as the only training points, take $r^*=4$ and remember that the intervals for AMS are represented by their left end points, then we get - since there is no overlapping of the basic functions from AMS - practically the same results from the finite element method and the AMS (dotted lines). If we use in addition the further point $f(s_3)=f(10\epsilon)=6$ we get with $r^*=4$ and the weight change formulas (8a,b) some disturbances for the formerly correct values at $f(s_1)$, $f(s_2)$ - full lines. For the finite element method an overlap is not allowed, so we have to reduce h to 2ϵ, doubling through this the necessary grid points, but avoiding errors at the $f(s_i)$ due to the forbidden overlap. For the also sketched $f(s)$ it cannot be said, however, that either of these two approximations represents this $f(s)$ any better, a usual statement for approximations of arbitrary nonlinear functions by any method being suitable in general for this purpose. But the AMS approximation saves an enormous amount of memory space, as has been shown in table 1, and has the further advantage, that the interpolated surface is constructed automatically from the scattered training points, since for linear interpolations like the method discussed one needs either a full training at equally spaced grid points or a search procedure for available points near the just considered stimulus. A respective method - MIAS - will be discussed later on.

II.4.2. Storage capacities

We will now turn to the question of how many different stimuli values can be stored in a memory with r places, using the simultaneous activation of a fixed number r^* of places, and how many different output values can be produced by our simple weight summation from the r^* activated places.

The amount of different stimuli, which can be stored is given by the value $\binom{r*}{r}$, which means the choice of r^* elements out of r available elements. A very conservative estimate which is much more easily calculated has been given in this connection by J. Albus in /1/. It can be derived in the following way:

$$(20) \quad \binom{r*}{r} = \frac{r!}{r^*!(r-r^*)!} = \frac{[r(r-1)...(r-r^*+1)](r-r^*)!}{r^*!(r-r^*)!}$$

$$> \frac{(r-r^*)^{r^*}}{r^*!} > \frac{(r-r^*)^{r^*}}{(r^*)^{r^*}}$$

$$= (\frac{r}{r^*} - 1)^{r^*}$$

So, if one has R different values for each of the n elements of the stimulus vector $\underline{s}$, one can safely store the outcoming R^n different stimuli possibilities, if $R^n < (\frac{r}{r^*} - 1)^{r^*}$. E. g. for four inputs and a memory with r = 10 000, $r^* =16$ one obtains from (20) an allowed resolution R for the inputs equal to R = $624^4 \approx 1{,}5 \cdot 10^{10}$, which is enormous.

In conclusion, one can say that in general no problems will exist with stimuli storage.

For the output side, it is helpful to define in addition to the characterization of trained weights by α_j - see formula (5b) - and of test points considered to contain enough information by $\tilde{a}_j$ - see formula (12) - the activated storage places by a_j^* (compare fig. 5) through:

$$(21) \quad a_j^* = \begin{cases} 1 & \text{for storage places } j\epsilon\{1,2...r\} \text{ activated} \\ & \text{by the coding procedure} \\ 0 & \text{otherwise} \end{cases}$$

For a certain stimulus $s^{(\kappa)}$ one obtains the respective output $p^{(\kappa)}$ - here for simplification we restrict ourselves again to one output of the respective equations - by superposition of the weights w_j being stored in the activated places. In the case where the memory is fully trained, all α_j from (5b) are equal to one and one may write e. g. equation (6a) by using equation (21):

$$(22) \quad p^{(\kappa)} = \frac{1}{r^*} \sum_{j=1}^{r} a_{j(\kappa)}^* w_j$$

The index (κ) in $a^*_{j(\kappa)}$ indicates, which of the a_j are 0 or 1 determined by the considered stimulus $s^{(\kappa)}$. If one assembles now all outputs $p^{(\kappa)}$ into a vector $\underline{p}$ one can write the weights w_j as a vector $\underline{w}$ of the dimension r and compress the $a^*_{j(\kappa)}$ into a matrix A^*, which has only zeros and r^* values $1/r^*$ in their rows, the position of the $1/r^*$ values being dependent of the stimulus value $s^{(\kappa)}$ and its respective encoding. This leads to the weight - output description:

$$(23) \qquad A^* \quad \underline{w} \quad = \underline{p}$$
$$ x \cdot r \quad r \cdot 1 \quad x \cdot 1$$

(due to X. Mao, unpublished, and also P. C. Parks and J. Militzer in /14/) in which the dimensions are indicated below the equation. Now, the maximum number of linearly independent equations is limited by r, the number of used storage places (e. g. 10 000 for each of the in general m outputs). So we cannot represent more than r prescribed output values $p^{(k)}$ exactly. If we ask for more specified output values than the number of linearly independent equations which exist, we obtain from our weight generating algorithm, however, a good approximation to the specified $p^{(k)}$- values as will be discussed in the next section "convergence".

The conclusion is that the response resolution, being limited by the number of available storage places in principle, is less freely variable than the stimulus resolution. But this has not turned out to present any difficulty in any of the examples and/or applications handled up to now.

II.4.3. Convergence

Theoretical proofs of convergence for the weight storage procedure described above were not given until quite recently. P. C. Parks solved this problem during a guest professorship at the TH Darmstadt in cooperation with J. Militzer; a detailed description of their results can be found in /14/. Here, only the general scheme of the proof and the major results will be presented. Some parallel work was done by D. Ellison /15/.

As a starting point we can use equation (23) with its background that by adjusting the weights in a number of say N training steps to generate x desired outputs we are dealing with the solution of linear equations. In principle, we know an appropriate solution to equation (23), independent of the number of equations and their consistency and/or inconsistency.

That is

$$(24) \qquad \underline{w} = (A^*)^+ \cdot \underline{p}$$

with $(A^*)^+$ being the Moore-Penrose-pseudoinverse of A^*, giving the exact inverse for a quadratic A^* with full rank, the minimum norm solution in the case that there are less linearly independent equations than necessary to define a unique $\underline{w}$ and the minimum of the squared solution errors $(A^*\underline{w} - \underline{p})^T(A^*\underline{w} - \underline{p})$ in the case that there are more linear independent equations than elements of $\underline{w}$, an overdetermined and/or inconsistent situation (see e. g. /16/). However, - as pointed out in /14/ - for the dimensions of A^* considered (thousands of rows and columns) the computation of $(A^*)^+$ is virtually impossible. The AMS algorithm on the contrary is easy to implement and exhibits the enormous advantage of computational simplicity: The operations involve only calculation of a scalar product of two rx1 vectors in which one of the vectors has only r^* non-zero elements. Additionally it is not necessary to store the previously calculated weights, they can be discarded, when the new weights are calculated (compare e. g. (31)). In fact, the convergence proof in /14/ shows explicitly for cyclic training - repeated training of the same $\tilde{N}$ stimulus situations passed through in the same order over and over again - that for $\lambda \cdot \tilde{N}$ training steps, with $\lambda \to \infty$, exactly the solution given by (24) is reached when the $A^* \underline{w} = \underline{p}$ equations are consistent.

Making again no distinction between trained and untrained points ($\alpha_j = 1$, $\tilde{r}^* = r^*$) we obtain from (8b) for example for all r^* weight positions that are activated by going from training step $(\kappa-1)$ to training step κ with the numbering $w_{j(\kappa)}$, $1 \leq 1(\kappa) \leq 2(\kappa) \ldots \leq r^*(\kappa) \leq r$, to characterize the activated weight positions:

$$(25) \quad \begin{bmatrix} w_{1(\kappa)} \\ w_{2(\kappa)} \\ \vdots \\ w_{r^*(\kappa)} \end{bmatrix} = \begin{bmatrix} w_{1(\kappa)} \\ w_{2(\kappa)} \\ \vdots \\ w_{r^*(\kappa)} \end{bmatrix}_{\kappa-1} + \frac{1}{r^*} \begin{bmatrix} r^* p^{(\kappa)} - w_{1(\kappa)} - w_{2(\kappa)} - \ldots - w_{r^*(\kappa)} \\ r^* p^{(\kappa)} - w_{1(\kappa)} - w_{2(\kappa)} - \ldots - w_{r^*(\kappa)} \\ \vdots \\ r^* p^{(\kappa)} - w_{1(\kappa)} - w_{2(\kappa)} - \ldots - w_{r^*(\kappa)} \end{bmatrix}_{\kappa-1}$$

Convergence of this algorithm would mean, that for $\kappa = 1,2\ldots N$, $N \to \infty$ a weight configuration is reached, from which by using the output calculation (6a) for all $p_i^{desired}$ the exact value is computed:

$$(26) \quad \frac{1}{r^*}\left[w_{1i}+w_{2i}+\ldots+w_{r^*i}\right] \to p_i^{desired}$$

We can now give (25) following /14/ a geometrical interpretation. Adding up the rows of (25) leads to:

$$(27) \qquad w_{1(\kappa)} + w_{2(\kappa)} + \ldots + w_{r^*(\kappa)} = r^* \cdot p^{(\kappa)}$$

Dividing (27) by $\sqrt{r^*}$ and turning to the r-dimensional vectorspace, we can embed (27) into:

$$(28) \qquad \tilde{\underline{a}}^{*T}_{(\kappa)} \, \underline{w} = \sqrt{r^*} \cdot p^{(\kappa)}$$

This describes as a linear equation a hyperplane in the r-dimensional space of $w_1, w_2 \ldots w_r$, the values $w_{j(\kappa)}$ specifying a special point on this hyperplane. Since for each pair of points $\underline{w}_\lambda$, $\underline{w}_\sigma$ out of (28) the following relation exists:

$$(29) \qquad \tilde{\underline{a}}^{*T}_{(\kappa)} \underline{w}_\lambda - \tilde{\underline{a}}^{*T}_{(\kappa)} \underline{w}_\sigma = \tilde{\underline{a}}^{*T}_{(\kappa)} [\underline{w}_\lambda - \underline{w}_\sigma] = 0$$

one finds, that the vector $\tilde{\underline{a}}^*_{(\kappa)}$ is orthogonal to all lines in the plane and by this is normal to the plane. By the division through $\sqrt{r^*}$ one gets furthermore, that $\tilde{\underline{a}}^*_{(\kappa)}$ having $r - r^*$ zero elements and r^* elements of the value $1/\sqrt{r^*}$ is a unit vector in r-space. Now, due to

$$(30) \qquad \begin{aligned} \tilde{\underline{a}}^{*T}_{(\kappa)} \, \underline{w} &= |\tilde{\underline{a}}^*_{(\kappa)}| \, |\underline{w}| \, \cos(\tilde{\underline{a}}^*_{(\kappa)}, \underline{w}) \\ &= |\underline{w}| \, \cos(\tilde{\underline{a}}^*_{(\kappa)}, \underline{w}) \\ &= |\underline{w}| \qquad\qquad\qquad \text{for } \underline{w} \text{ parallel to } \tilde{\underline{a}}^*_{(\kappa)} \\ &\overset{(28)}{=} \sqrt{r^*} \cdot p^{(\kappa)} \end{aligned}$$

$|\underline{w}| = \sqrt{r^*} \cdot p^{(\kappa)}$ applies for $\underline{w}$ having the direction of the normal to the hyperplane and this is just the case for $\underline{w}$ being the perpendicular from the origin to the hyperplane (compare fig. 19a).

With vectors $\underline{w}$ and $\tilde{\underline{a}}^*$ in r space the algorithm (25) may now be written:

$$(31a) \qquad \underline{w}_\kappa = \underline{w}_{\kappa-1} + \mu_\kappa \tilde{\underline{a}}^*_{(\kappa)}$$

$$(31b) \qquad \mu_\kappa = \sqrt{r^*} \cdot p^{(\kappa)} - \tilde{\underline{a}}^{*T}_{(\kappa)} \underline{w}_{\kappa-1}$$

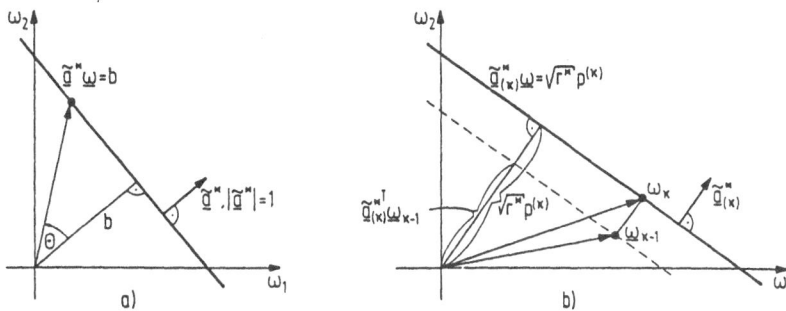

<u>Fig. 19:</u> a) Hyperplane, normal to $\tilde{\underline{a}}^*$ with distance b from origin to hyperplane (r = 2, $\Theta = \cos(\tilde{\underline{a}}^*,\underline{w})$);   b) Geometrical interpretation of weight updating as constructing the perpendicular from point $\underline{w}_{\kappa-1}$ in r-space to the hyperplane generated in step κ.

Actually by (31a) only the weights in the activated storage cells as characterized by the r^* non-zero values of $\tilde{\underline{a}}^*_{(\kappa)}$ are changed, the amount of change being given in accordance with (25) by μ_{κ}. The geometrical interpretation as put forward in /14/ can be derived from fig. (19b): One goes from $\underline{w}_{\kappa-1}$, the result of the $(\kappa-1)$th training step, in the direction of the normal to the hyperplane presenting the result of the κth training step just by the difference between the distance of the hyperplane from the origin and a hyperplane parallel to the considered hyperplane through $\underline{w}_{\kappa-1}$. This is equivalent to constructing the perpendicular from the point $\underline{w}_{\kappa-1}$ to the hyperplane characterized by the index κ, so that (31) can be interpreted as always constructing the perpendicular onto the next hyperplane and using its footpoint as the next point in the r-dimensional weight-space.

We can now combine all the equations (28) for the N training steps into one matrix equation:

$$(32) \qquad \tilde{A}^* \underline{w} = \sqrt{\underline{r}^*} \cdot \underline{p}$$

which is essentially the same as (23), but has the non-zero-elements $1/\sqrt{r^*}$ in $\tilde{A}^*$ instead of $1/r^*$ in A^*, which is the reason the additional tilde in this section has been used.

Let us assume, that (32) is consistent, that means, that all hyperplanes intersect in at least one point $\underline{w}^*$, which is by this a solution to (32). Considering the distance between this point $\underline{w}^*$ and the points $\underline{w}_{\kappa}$ generated in our weight updating scheme, we get for the squared distance:

$$(33) \qquad V_{\kappa} = (\underline{w}^* - \underline{w}_{\kappa})^T (\underline{w}^* - \underline{w}_{\kappa})$$

the rate of change:

$$(34) \quad V_\kappa - V_{\kappa-1} \quad = (\underline{w}^* - \underline{w}_\kappa)^T (\underline{w}^* - \underline{w}_\kappa) - (\underline{w}^* - \underline{w}_{\kappa-1})^T (\underline{w}^* - \underline{w}_{\kappa-1})$$

$$\overset{(31a)}{=} \quad (\underline{w}^* - w_\kappa)^T (\underline{w}^* - w_\kappa) - (\underline{w}^* - \underline{w}_\kappa + \mu_\kappa \tilde{\underline{a}}^*_{(\kappa)})^T (\underline{w}^* - \underline{w}_\kappa + \mu_\kappa \tilde{\underline{a}}^*_{(\kappa)})$$

$$= (\underline{w}^* - \underline{w}_\kappa)^T (\underline{w}^* - \underline{w}_\kappa) - (\underline{w}^* - \underline{w}_\kappa)^T (\underline{w}^* - \underline{w}_\kappa) -$$

$$- 2\mu_\kappa \bullet (\underline{w}^* - \underline{w}_\kappa)^T \tilde{\underline{a}}^*_{(\kappa)} - \mu_\kappa^2 \tilde{\underline{a}}^{*T}_{(\kappa)} \tilde{\underline{a}}^*_{(\kappa)}$$

$$\overset{(*)}{=} \quad - \mu_\kappa^2$$

For the reduction (*) we have used on one hand that $\tilde{\underline{a}}^*_{(\kappa)}$ is a unit vector so that $\tilde{\underline{a}}^{*T}_{(\kappa)} \tilde{\underline{a}}^*_{(\kappa)} = 1$, and on the other hand that $\underline{w}^*$ lies as a point of intersection of all hyperplanes as well in the plane defined by (28) as $\underline{w}_\kappa$ does , so that $(\underline{w}^* - \underline{w}_\kappa)$ lies in this plane, for which $\tilde{\underline{a}}^*_{(\kappa)}$ is the normal, so that $(\underline{w}^* - \underline{w}_\kappa)^T \bullet \tilde{\underline{a}}_{(\kappa)} = 0$.

From (34) it follows, that the distance of the points $\underline{w}^*$ decreases monotonically, so that for $\kappa \to \infty$ $\underline{w}^*$ has to be reached finally. (Thus V_κ is a Liapunov function). This behaviour is also to be expected from our geometrical interpretation without the above mathematical reasoning: since $\underline{w}^*$, $\underline{w}_\kappa$ are both in the same hyperplane and $\underline{w}_\kappa$ is found from $\underline{w}_{\kappa-1}$ by constructing the perpendicular from $\underline{w}_{\kappa-1}$ onto this hyperplance, $\underline{w}^*$, $\underline{w}_\kappa$, $\underline{w}_{\kappa-1}$ are forming a triangle in r-space with a right-angle at $\underline{w}_\kappa$. Since $|\underline{w}_{\kappa-1} - \underline{w}^*|$ forms the hypothenuse of this rectangular triangle it is always greater than the other sides, one of which is formed by $|\underline{w}_\kappa - \underline{w}^*|$. So also by this argument the distance $|\underline{w}_\kappa - \underline{w}^*|$ is decreasing monotonically.

The rate of convergence is, by the way, rather slow and there may exist schemes of improvement. Some such schemes are investigated in Parks and Militzer /17/. However, one pays for this by an increase on the complexity of the algorithms and one has to weigh this against overall numerical effectiveness especially since one can live in general - at least in control loops - very well with just "fairly good" process/nonlinear characteristics approximations.

All the arguments put forward up to now on convergence properties are independent of the training procedure - cyclic or random - and of the starting point. However, the training points (N) actually used may be less than the amount of available storage places (r). Then the vectors $\tilde{\underline{a}}^*_{(\kappa)}$ do not span the whole r-dimensional space (underdetermined case of (32)), so some orthogonal, not reachable subspace will exist. It is at least plausibel, that the components of $\underline{w}^*$, which are lying in this subspace, are not determined by the training procedure: $\overset{*}{\underline{w}}$ is not uniquely specified. Actually the $\overset{*}{\underline{w}}$ to be reached is dependent of the starting point of the training procedure. As shown in /14/ those components of the starting point $\underline{w}_0$ (initial values in the memory), which lie in the subspace orthogonal to the subspace spanned by the $\overset{\sim *}{\underline{a}}_{(\kappa)}$, determine the components of $\overset{*}{\underline{w}}$ out of this subspace: The respective components of $\underline{w}_0$ are not affected by the training procedure, as one would assume anyhow.

A final question left open in the preceding discussion is, what happens, if (32) is inconsistent. Then no $\overset{*}{\underline{w}}$ will exist. In case of cyclic training it is proven in /14/, that the point $\overset{*}{\underline{w}}$ will be replaced by a limit cycle of points. In the case of random training up to now only a conjecture from P. C. Parks exists, that the algorithm will tend to reach some "minimal capture zone" with a high probability. The size of the limit cycle and of the minimal capture zone are determined by the amount of inconsistency, that means the size of the minimal simplex, which is included by the intersecting hyperlanes.

It should be pointed out, finally, that by the geometric interpretation of the Albus scheme, Parks and Militzer (/14/) established also a connection to already existing mathematical procedures for solving systems of linear equations. Actually, the Albus scheme is practically identical to the Karczmarz algorithm /19/ and one has the advantage by this statement, that further work on this algorithm, like the work of Aved'yan /20/ can be used for convergence acceleration investigations (see /21/).

II.5. Variable Generalization

II.5.1. Principal Considerations

Up to now the area of generalization and consequently of interpolation was a fixed, fundamental parameter of the memory.

What does that mean? Let us clarify at first, which properties of the memory are not influenced by this parameter value: the general fineness for structuring the nonlinear manifold above the input space is influenced only by the basic quantizations ϵ_i [4] on one hand and by the number of different output values - which was shown to depend on the number r of available memory cells - on the other hand, but not by the size of generalization $\overset{*}{r}$. However, the amount of effort for reaching the finest possible differentiation grows rapidly with $\overset{*}{r}$. This is due to that fact that a high $\overset{*}{r}$ means, that any modification of the trained value at a certain stimulus point changes the values at points

---

[4]For simplification $\epsilon_1 = \epsilon_2 = ... = \epsilon_n$ and so on is used in the following argumentation.

in the whole distance covered by r^*, making already correct responses incorrect to some extent. So the training effort to get different values for nearby stimuli exactly is much greater for large r^* than for small r^*.

Now, a large ϵr^* has the advantage of giving very quickly at least some answers with very few training points, which is important for reaching a first rough orientation in unknown, new situations. Also such a rough orientation is in general sufficient for stimuli situations seldomly met, since much better answers are required for situations frequently met and being important therefore for the process and/or individual. This leads to the following strategy as being advantageous: At the commencement of learning a large generalization ϵr^* is used, which will be reduced lateron in areas well covered. Such a strategy will be called "variable generalization".

Two approaches for variable generalization are discussed in this chapter: at first some implementation of it using the AMS concept will be described. This is based on a stepwise reduction of the generalization area with relatively big differences in the ϵr^*-values. Then some other approach will be put forward, in which no reference is made to ideas about neuronal information processing, but just concepts from scattered data interpolation out of the area of computer graphics are used, leading to a smooth reduction of ϵr^* (/22/). Finally a short evaluation of advantages and disadvantages of both concepts will be given.

II.5.2. Concept of variable generalisation using the AMS

The degree of generalization should be data driven, that means automatically dependent on the denseness of stimuli used for learning/training. A very simple implementation of this idea is to use a number of AMS memories, say k, with different degrees of generalization $(\epsilon r^*)_{(j)}$; $j = 1,2,...k$. In the training cycle all memories are taught the stimulus-response connections in parallel - see fig. 20 -. During recall it is checked in the memories, beginning with $(\epsilon r^*)_{min}$, whether, for the given stimulus, $\eta \geq \eta_{min}$ is fulfilled. The first memory from the chain of memories with growing ϵr^* meeting this condition is selected, to give the output vector $\underline{p}$.

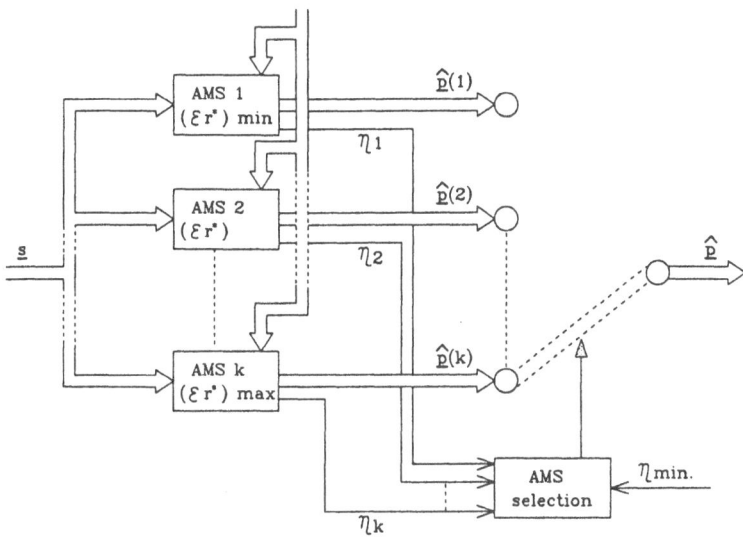

Fig. 20 (from /23/): Variable generalization scheme in connection with AMS. The selection box connects the output with that memory with smallest ϵr^*, which fulfills $\eta \geq \eta_{min}$. During training, all memories are filled with the stimulus-response pairs in parallel.

If there is a permanent alternation between training of further stimuli-response pairs and recall of responses for certain stimuli - as in the case of the learning control loop from fig. I.5 - one would find for the process model being stored in a number of memories, a behaviour like that shown in fig. 21. The details of the respective control task and learning control loop are not explained here. Only the general behaviour of the memory chain response will be demonstrated: In the beginning no response is found in any of the memories for the stimulus of interest; then just for one stimulus a response with $\eta \geq \eta_{min}$ is found in the memory with $r^* = 16$, the maximum generalization in the considered case ($\epsilon = 1$). After a while a second and third success is achieved with that memory. Then the AMS with $r^* = 8$ starts to give responses and one can see, that in case that no responses with $\eta \geq \eta_{min}$ in this memory can be found, now responses from the AMS with $r^* = 16$ always exist. Then the AMS with $r^* = 4$ starts to take over and finally in addition sometimes respones from a simple look-up-table (no generalization, $r^* = 1$) are supplied. The look-up-table gives, for exact training and no noise, the most accurate response, since there is no possibility of information distortion by points trained lateron.

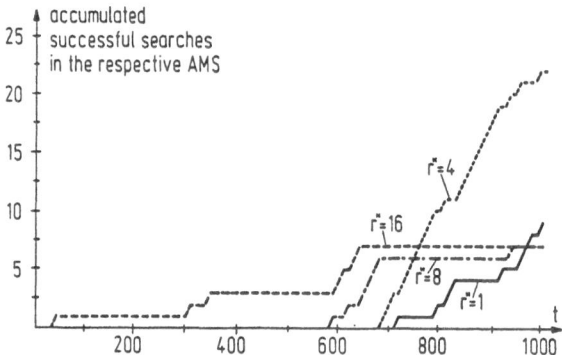

Fig. 21: Number of successful - $\eta \geq \eta_{min}$ - memory answers (accumulated) from four memories with $\epsilon = 1$, $r^* = 16$, 8, 4 and 1 (look-up-table), drawn up over a control loop running time, not explained here, since only the principle behaviour shall be demonstrated. Details of the respective control task are given in /23/, /24/ and section III.2.2.4.

As stated in II.5.1, the amount of generalization supplied by this implementation varies in a graded way with levels to be fixed in advance. Furthermore, the basic principle applied relies on the availability of additional memory space.

II.5.3. Mathematical alternative MIAS

The problem of interpolation between randomly distributed and/or scattered data does not arise only in learning, but also in the field of construction of smooth surfaces in the context of computer graphics. In 1986, J. Militzer devised a memory system with a smoothly varying amount of variable generalization by combining ideas from D.H. McLain 1974 (/25/) and from R.J. Renka/A.K. Cline 1984 (/26/) from this field of mathematics. He called the result MIAS - McLain-type Interpolating Associative Memory - and this approach will be described now.

The first step in MIAS is to generate for a given stimulus $\underline{s}$ the distances d_i to all trained stimuli $\hat{\underline{s}}_i$ and to arrange those distances by re-numbering in their order of magnitude, so that one gets:

(35) $d_1(\underline{s}) \equiv |\underline{s} - \hat{\underline{s}}_1| \leq d_2(\underline{s}) \leq d_3(\underline{s}) \ldots$

As the influence area for the generalization the n-dimensional hypersphere having just $\hat{\underline{s}}_{q+1}$ as an element of its surface, that means with the radius $\rho(\underline{s}) = d_{q+1}(\underline{s})$, is now used. q is a free parameter of the memory system. Since the radius ρ is dependent on the trained data themselves, the envisaged adaptation to the density of training points is reached. Fig. 22 illustrates this concept for the case of two inputs and q=6.

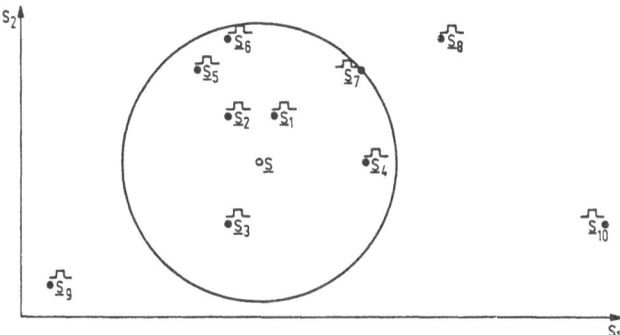

<u>Fig. 22:</u> Example for scattered data numeration with respect to a considered value $\underline{s}$ in the two-dimensional s_1, s_2-stimulus-space and for selection of the influence/generalization area by putting a hypersphere - here circle - through $\hat{s}_{q+1}$, here with q=6 through $\hat{s}_7$.

For the estimation $\hat{p}_i$ of the value of the output-vector element p_i a balancing hyperplane of the dimension n is considered, which can be written by suppressing again the index i:

$$(36) \qquad g(\underline{x}) = a_0 + \sum_{k=1}^{n} a_k x_k$$

the balancing being achieved by calculation of the coefficient-vector $\underline{a}$ via the performance-criterion:

$$(37) \qquad J = \sum_{l=1}^{q} w_l[g[\hat{s}_l - \underline{s}] - \hat{p}_l] = \overset{Min}{\underline{a}}$$

with the weighting factors

$$(38) \qquad w_l = [1/d_l(\underline{s}) - 1/\rho(\underline{s})]^2$$

which take training values nearer to $\underline{s}$ more heavily into account than more distant values (compare fig. 23).

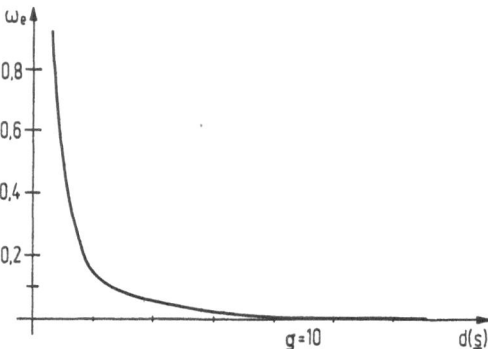

<u>Fig. 23:</u> Sketch of the weighting factor w_1 from (38) for q=10.

The wanted estimate $\hat{p}_i = \hat{p}$ on account of (36) is given by

(39) $\hat{p} = g(\underline{0}) = a_0$

Although MIAS gives responses for any point $\underline{s}$ in the stimulus space if only more than q points are trained, it may be that such responses are not helpful, since a balanced linear interpolation can be meaningless, if the response is requested too far from the known stimulus-response pairs (training points). As a solution one may define some minimal area for the neighbourhood of the considered stimulus $\underline{s}$, in which q points have to be found to regard $\underline{s}$ as a point with sufficient training in its environment and by this the estimated response vector $\hat{p}$ is considered to be reliable. Since this again is a training indication, but with a different definition than for AMS, we shall use here the expressions $\tilde{\eta}$, $\tilde{\eta}_{min}$ instead of η, η_{min} used in connection with the AMS. For calculating $\tilde{\eta}$ - being required to be greater or equal to a somehow fixed $\tilde{\eta}_{min}$ - no unique way can be derived from the MIAS approach itself. However, the following formula has proven to be suitable:

(40) $\tilde{\eta} = \dfrac{200}{200 + \rho(\underline{s})}$

and therefore has been always applied up to now.

The efficiency of MIAS was evaluated by using it on the one hand for the storage of the test functions FSIN, FCOS, CORNER form section II.3.1. (/27/) and on the other hand for the learning control loop form chapter I.4. (/22/). We shall deal in this section with the function storage problem only, since some results in connection with the learning control loop will be presented in chapter III.

As a first step in application of MIAS one has to select a value for the parameter q. An indication of what may be an appropriate choice can be gained from fig. 24, in which the quality criteria E_{bm} - see (13) - and E_{max} - see (15) - are plotted against q for the test function FSIN - see (16) -.

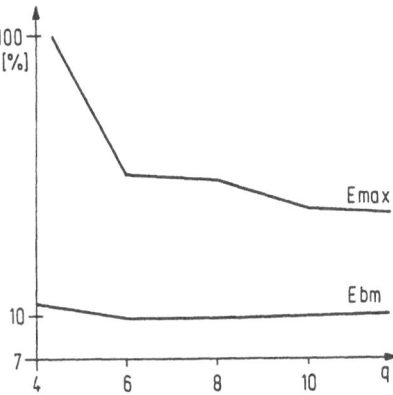

Fig. 24: Values of E_{bm} and E_{max} with respect to the number q of points used for interpolation, calculated for the test function FSIN.

The conclusion from this figure is that values greater than q = 5 seem to be necessary and values smaller than q = 11 sufficient, so that $6 \leq q \leq 10$ may be a suitable range for points to be taken into account for local interpolation/generalization.

Out of tests with FSIN, FCOS and CORNER fig. 25a demonstrates with CORNER used as example the general result, that MIAS (q=10) already reaches with 300 test points roughly the same quality as AMS gives with 1000 test points - fig. 16d -. Fig. 25b shows in addition, how much poorer a reproduction of CORNER with AMS and the values $\epsilon_1 = \epsilon_2 = 16$, $r^* = 16$ from fig. 16d would be, if only the number of training points used for MIAS, that means 300, would be taken into account.

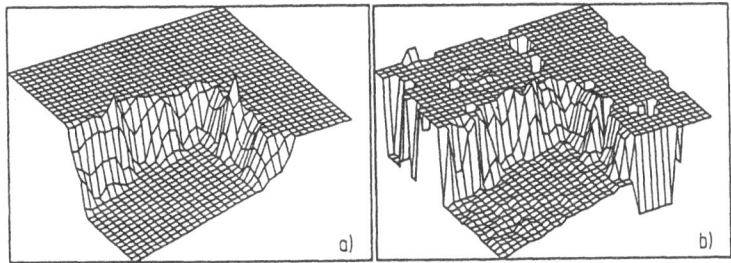

Fig. 25: Reproduction of CORNER from 300 training points by using in a) MIAS with q=10 and in b) AMS with $\epsilon_1 = \epsilon_2 = 16$, $r^* = 16$.

Table 9 puts forward some additional information about the relative quality to be reached with MIAS and 300 training points against AMS with 1 000 training points, now stretching over all three test functions. It has to be compared with respect to AMS with table 6. However, the main purpose in presenting table 9 is to show the effect of limiting the allowed interpolation/generalization area by introducing some $\tilde{\eta}_{min}$.

	$\tilde{\eta}_{min}$	E_{bm}	E_{max}	untrained test points
FSIN	-	9,9 %	30,7 %	-
	0,3%	6,5 %	18,0 %	66 %
FCOS	-	19,5 %	210,1 %	-
	0,3%	10,1 %	63,5 %	66 %
CORNER	-	3,0 %	80,7 %	-
	0,3%	3,0 %	70,5 %	66 %

Table 9 - Influence of limiting the area, in which training points must ly to be accepted for interpolation - MIAS, q=8, 300 training points.

One sees immediately, that by introducing $\tilde{\eta}_{min}$ = 0,3 the number of test points considered to have enough training points in their immediate neighbourhood is heavily reduced for the - relatively small - number of 300 training points (for definitions of test points etc. see section II.3.1). This had to be expected. The main result of table 9 is, however, the different extent, to which the error measures are changed by $\tilde{\eta}_{min}$ = 0,3: the highest effect can be observed for FCOS - especially on E_{max} - which indicates, that the setting up of some $\tilde{\eta}_{min}$ in MIAS is required at least in those cases, in which strong curvatures occur in the functions to be stored, since the degree of curvature is the main distinction between FCOS, FSIN and CORNER.

II.5.4. Comparison between AMS and MIAS

Some results regarding a comparison between AMS and MIAS have been given already in the last chapter. However a fair discussion on advantages and disadvantages has to be more systematic on the one hand and has to consider on the other hand further aspects, like:

- the necessary amount of software for system organisation

- the required computing time for response generation

- special properties

Table 10 supplements the hints already given by some further information on reproduction of FSIN
by MIAS and AMS. Since we have concluded from table 9, that the selection of $\tilde{\eta}_{min}$ is of some
importance for functions with a certain amount of curvature a way has to be found at first to make
the results of selecting η_{min} in AMS and $\tilde{\eta}_{min}$ in MIAS at least roughly equal. Following a
proposal of J. Militzer we can do this by selecting η_{min} and $\tilde{\eta}_{min}$ in a manner, by which nearly the
same percentage of test points considered untrained are produced by both memory systems. This is
the basis in the comparison of table 10. Furthermore in addition to a fixed generalization of AMS a
three staged variable generalization approach for AMS was included and finally two different
amounts of training points were taken into account.

FSIN	number of training points	E_{bm}	E_{max}	untrained test points
AMS $\rho \bullet \epsilon = 256$	300 1 000	14,4 % 5,7 %	29,4 % 11,3 %	12,5 % 2,4 %
AMS $\rho \bullet \epsilon = 128/256/512$	300 1 000	11,6 % 5,7 %	14,7 % 9,3 %	13,1 % 2,9 %
MIAS $q = 10$	300 1 000	9,4 % 2,9 %	22,5 % 11,3 %	11,7 % 2,8 %

Table 10 - Comparison of AMS with fixed generalization, variable generalization and MIAS for
FSIN. The effects of selecting η_{min}, $\tilde{\eta}_{min}$ are equalized by considering cases with roughly the same
amount of untrained test points.

One sees from table 10, that regarding the mean absolute error E_{bm} MIAS gives always the better
results, however, the staged variable generalization does not seem to have a real advantage
vis-à-vis the fixed generalization for a satisfactory degree of training (1 000 training points). This is
due to the fact that for the function storage problem there exist no areas with a higher amount of
training and with a lower amount of training after a sufficient number of stimulus-response training
sets have been generated by the used random distribution over the stimulus area. However, a
certain advantage is present, if the interpolation is adapted directly to the density of trained points,
as is shown by the better results from MIAS. This conclusion leads to a certain confusion if one
takes a look on the maximum absolute errors E_{max}. Here surprisingly, the staged AMS is not only
better than AMS with fixed generalization but also than MIAS with its generalization fully adapted
to training point density. Why this happens can be deduced from fig. 26. Since for estimating the
response value AMS takes into account only trained values from points for which the considered
point is in the middle, this is not the case for MIAS. With trained points inappropriately
distributed even an extrapolation instead of an interpolation may take place, leading to rather poor
response estimations. This influences the maximum absolute error directly, but - since normally
being a rare case - not to a measurable extent the mean absolute error.

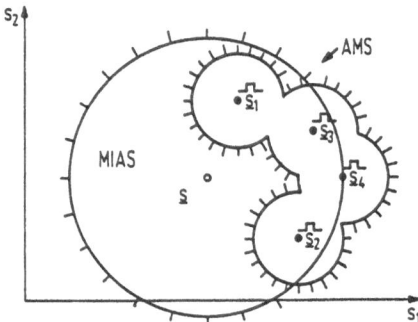

Fig. 26: General sketch indicating, that MIAS (here q = 3 for simplicity) may interpolate and/or extrapolate very keenly in case of inappropriate distribution of trained points, since AMS interpolates only and generalizes by this more conservatively.

Another point less favorable to MIAS is, that it is missing the possibility for direct noise filtering, which was pointed out for AMS as a result of memory place sharing by generalization areas due to distributed data storage. However, the influence of noise is indirectly reduced by building up hyperplanes out of more than the minimally necessary points with the balancing procedure (37).

From the point of view of necessary software for system organisation, MIAS requires roughly a 1,5 fold program size as compared to AMS, which is normally no problem.

For the response generation from a certain stimulus $\underline{s}$, the required time for AMS is dependent on the dimension n of $\underline{s}$ and the degree r^* of generalization (compare table 2), but independent on the amount of trained values. Roughly one gets for it 5-8 msec for $n \leq 4$, $r^* = 16$ with a PDP 11/34 as pointed out earlier.

MIAS is somewhat slower, since there is a basic computation time dependent on the stimulus dimension n and the number of points considered q, which lies for n = 2, q = 8-10 already in the size of 5-8 msec for the same machine installation and program type as described for AMS, and to which furthermore an additional time, dependent on the number of training points, has to be added. This dependence of the number of trained stimulus-response pairs is due to the fact that all distances between the considered point and the trained stimuli have to be calculated before a decision is possible, which training points have to be taken into account for the interpolation. For each training point an additional amount of 0,04 msec has to be added. However this linear dependence can be reduced to a logarithmic dependence by a hierarchical focussing approach as described by S.Y. Oh in /28/.

It has to be remarked finally, that less effort has been put into optimizing MIAS than into optimizing AMS up to now, so that the cited time and program size relations may still change slightly to the advantage of MIAS.

In general regarding variable generalization one should keep in mind that its major merits emerge in problems, where cases of different degrees of training are met simultaneously, as in dynamic systems, where transitions between different set points supply much less data than the effort to keep a certain set point. That means, the real power of variable generalization manifests itself in the learning control loops to be discussed in detail in the next chapter and this topic will be addressed there again, therefore.

II.6. Direct application areas

AMS and MIAS can be considered as examples of a class of memory systems, which is characterized by being able to generate good output value estimations in a m-dimensional output vector space by just a decent training with input-output relations from irregularly distributed n-dimensional input vectors. Such automatically interpolating memory systems have - besides the learning control loops to be dealt with in the next section - also very important direct application areas such as:

1.) Representation of n-dimensional (nonlinear) characteristic fields of technical processes.

Fig. 27 shows the idea: If one uses the inputs of test runs as input stimuli and the respective output values as responses, one can after a sufficient training phase with a locally generalizing memory run in parallel take it away from the test facility and use it as a (nonlinear) process-simulator for process handling and/or control studies.

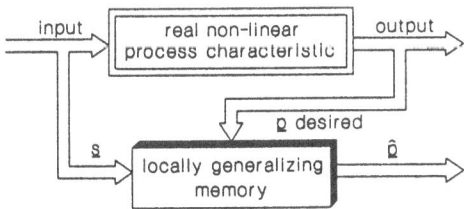

Fig. 27: Memorization of process characteristics by letting a locally generalizing memory run in parallel to tests with the process results in a process-simulator

2.) Inversion of n-dimensional (nonlinear) characteristic fields of technical processes.

If one changes the associative memory inputs and outputs in fig. 27 according to fig. 28, the memory system learns to deliver the appropriate process inputs for generating the process outputs met during the testing phase. Such an inversion of the (nonlinear) process characteristics can be put in front of the real process input as precompensator and for additional feedforward control input, leading to a high reduction of nonlinear behaviour and/or input-output coupling, and by this means to simplifications of feedback control loop synthesis. As long as unique relationships are learned it is not necessary to make assumptions like equal number of inputs and outputs or general invertibility. But there exist problems, in which one has to be careful as in the case of robot end effector positions. Here the same output (end effector position/orientation) can be reached with different inputs (robot joint angles), so that additional constraints have to be taken into account (learned) to get the inversion working.

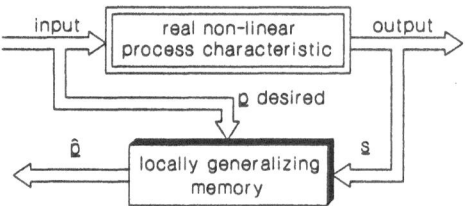

Fig. 28: Use of the memory system to learn the inverse process characteristics, leading to decoupling and linearization if the system is used later on as a precompensator for the process.

The characteristic fields just considered are time independent. However dynamic problems like non-linear process-behaviour and/or signals may also be learnt. The basic principle is to span the input space not only by the input values, but also by some history for each input value. That means, the stimulus $\underline{s}$ reads $(s_1(k \cdot T), s_1[(k-1)T], s_1[(k-2)T], ..., s_2(kT), s_2[(k-1)T], s_2[(k-2)T] ...; s_i(kT), s_i[(k-1)T], s_i[(k-2)T] ...)$, T being the sampling time. In the blockdiagram this stimulus extension can be represented by a short term memory.

An important example for a direct application of the automatically interpolating memories in the field of dynamic systems is

3.) "Automatic transfer of operator strategies"

If a process is run by an operator his strategies can be learned by a locally generalizing memory with the possibility of replacing the operator later on. For this purpose one feeds the memory with the process-inputs and -outputs including eventually some history of these signals to characterize the actual situation (see fig. 29). From the process inputs the actual value is treated as memory response to be learned. If the situation characterization is including all information the operator is using, the operator can safely walk away after a sufficient training time. If one considers the efforts to substitute human operators by artificial intelligence systems with their necessity to extract the strategy from the human operator verbally by specialized knowledge engineers, the method proposed here is remarkably simple

We tested it in our laboratory by simulating a chemical reactor with two inputs and two outputs on a computer, the control interface being a separate box with two knobs to turn. The process responses were shown on a graphical display. Students, who did not know the process and its non-linear characteristics were asked to learn a satisfactory set-point control strategy just by trial and error. The memory was fed with process inputs/outputs and students control measures in parallel just as explained above. The results were very good. When the students had learnt to control the process, the associative memory system (AMS) could do it also, partly even better due to the smoothing effect on actions stemming from local generalization. (For results on respective tests see fig. III.17 and fig. III.30). Some similar experience is reported in /29/ for the completely different field of car driving strategies. For real applications, however, one would have to be careful that all possibly occurring situations are met during learning. Eventually some safety backup solutions for handling of not expected/trained situations have to be installed at least, if the process is hazardous. But this has to be the case for running the process with human operators, too.

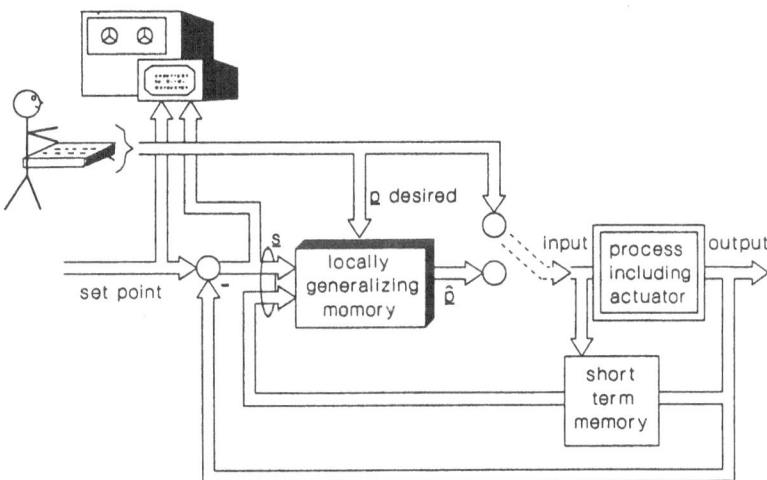

Fig. 29: Set up for automatic transfer of operator strategy into a locally generalizing memory.

It should be stressed finally, that the important feature for all three application areas described is the generalization/interpolation capacity of the memory system. Only this allows one to live with irregularly distributed training inputs and a finite amount of training.

II.7. Comparison with "neural nets" and "polynomial approximation"

A very popular approach to learning systems is the so-called "neural net" as a general model of neural information processing implementation in living creatures. There exist different formulations of this idea. A recent review, in which details are explained, has been given by Lippmann in /30/. We shall restrict ourselves here to the most simple form of a three layered net, with just one layer (the "hidden" layer) between the input neurons and the output neurons - see fig. 30a -.

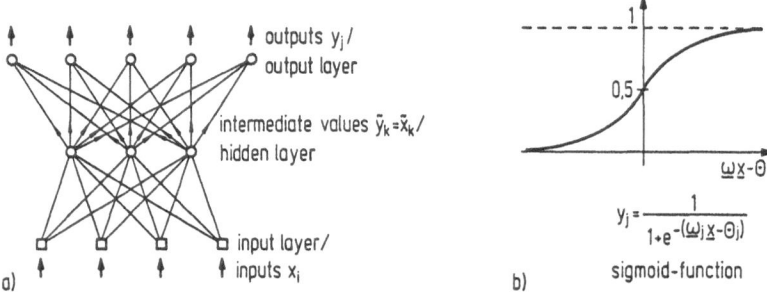

Fig. 30: Neural net (a)) and sigmoid-function (b)) as threshold in the neural net

In the neural net considered the information is fed forward from the inputs to the outputs by a direct connection of the neurons from a certain layer to all neurons in the next layer. The incoming information is multiplied by adjustable weights w_{jk} - responsible for the learning capacity of the net - summed up and then processed by a nonlinearity, normally a threshold or something similar, e. g. the sigmoid function shown in fig. 30b.

Neural nets can be used for both classification tasks as well as for interpolation tasks. For interpolation, in which we are interested here, it is important to note that, by using a neural net of the form shown in fig. 30 with at least one hidden layer, any considered non-linear multi-input/multi-output relationship $\underline{y} = \underline{f}(\underline{x})$ can be approximated with any desired precision. This fact is based on a theorem by Kolmogorov /31/ and has been worked out by Lorentz in /32/ and Hecht-Nielsen in /33/.

The performance of neural nets is to a great extent dependent on the number of hidden layers, the number of neurons in the hidden layers, the nonlinearities used for the characterization of the neurons and the training algorithm (learning and/or adaptation rule) applied. Indeed, applying the theorem of Kolmogorov, one can conclude that for n input variables $2n + 1$ neurons in just one hidden layer are sufficient to approximate any given function by given nonlinearities with one input, but the form of these nonlinearities is not known (other than being monotonic in their arguments). Using freely chosen nonlinearities, like the sigmoid-functions from fig. 30b, one may compensate for the missing adaptation to the function to be approximated by implementation of more than the just necessary amount of neurons and eventually additional hidden layers. However, if not enough neurons are provided, the multi-input/multi-output relationship cannot be approximated well enough and if too many neurons are implemented the training effort may be needlessly high. So the appropriate number of neurons for a given problem has to be found by some search procedure with respect to the detailed construction of the neural net in general.

For comparing the performance of neural nets with AMS, M. Hormel from my group used the simple net of fig. 30, the two input, one output relationship $y = f(x_1, x_2) = \sin^2 x_1 \cdot \sin^2 x_2$ with $0 \leq x_1 \leq \pi$, $0 \leq x_2 \leq \pi$, 5 neurons in the hidden layer (following tests with at first a higher number of neurons) using the "back propagation" training algorithm. Back propagation is a gradient calculation procedure (see e. g. /30/) by which the mean square errors between the actual and the desired output are minimized in an iteration procedure through appropriate changes of the weights w_{ij} in the network: it is considered to be a fast, good algorithm for neural net training. Fig. 31 leads to the following conclusion: AMS gives a satisfactory result with much fewer training steps than the neural net. Furthermore, intermediate results already show a fair approximation to the given function using AMS, whereas with the neural net approach this is not the case.

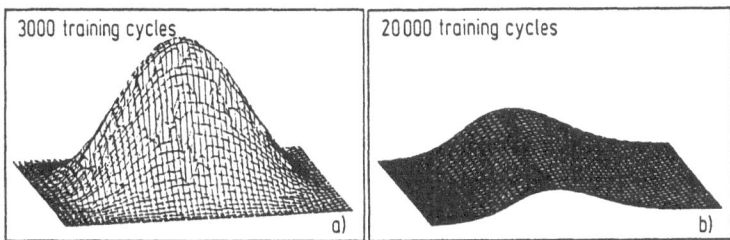

Fig. 31: (due to M. Hormel, W. Heinke) Approximation of $y = \sin^2 x_2 \cdot \sin^2 x_2$, $0 \leq x_1, x_2 \leq \pi$ with AMS, $r^* = 16$ - a), and with a neural net shaped according to fig. 30 with 5 neurons in the hidden layer and back propagation as training algorithm - b), 1 training cycle = 1 randomly chosen training point.

The reason for this different behaviour is, that in the neural net we are trying to approximate the given function not by approximation values only locally defined but by globally defined functions in $0 \leq x_1, x_2 \leq \pi$ [5]. Clearly one knows from mathematical theory, that any quadratic integrable function can be approximated to any desired degree of precision by an appropriate superposition of a set of linear independent quadratic integrable functions. However, the effectiveness of this approximation is highly influenced by the selection of the basic functions. In section II.4.1. we have discussed that the AMS approach is similar to the finite element approach for constructing numerical approximations to the solution of differential equations. Now, the alternative to the finite element approach in solving partial differential equations approximately is the superposition of the

[5] A similar statement - however, supported by other arguments/examples - that local approximation is preferable to and much more effective than global approximation by the general neural net approach can be found in /34/, which was brought to my attention by P. Parks in February 1991.

eigenfunctions of the considered problem. This is in general a very effective approximation - good approximation of the exact solution with only very few basic functions - but it relies, however, on problem-dependent basic functions, which are shaped not only by the considered differential equations but also by their boundary conditions. Due to this fact one should work with the locally approximating finite element method, and not with the globally defined eigenfunctions, as long as one is interested in a flexible tool for a numerical, problem-independent solution of partial differential equations.

If one accepts that AMS is a model of the cerebellum, one could argue that in the evolution of live a locally approximating approach turned out to be preferable to a general, globally approximating neural net approach. So there also exist theoretical reasons to use the more structured AMS instead of the general neural nets in interpolation problems in addition to the practical results from fig. 31 for a special comparison.

The argument for a refinement of the neuronal structures to obtain better solutions for special task achieved during evolution of life seems, by the way, also to apply to the classification problem. In /35/ it is shown, that for the classification task the more structured "feature map" neural model from Kohonen et. al. (/36/) achieves the respective results much faster than general neural net approaches.

Finally, the big advantages of easy implementation of neural net structures such as parallel processing units, giving fast multipurpose hardware, applies also for AMS and the feature map.

As with AMS and MIAS, there exist also mathematical alternatives to the neural net, if we restrict ourselves to its basic feature of general function approximation by appropriate superposition of globally defined given functions. A respective tool is the superposition of polynomials of different degree:

$$(41) \qquad y = \beta_0 + \sum_i \beta_i x_i + \sum_i \sum_j \beta_{ij} x_i x_j + ... \sum_i ... \sum_q \beta_{i...q} \, x_i ... x_q$$

The main problem here is to select a manageable number of elements from (41) which guarantees a good approximation. Respective automatic approaches can be found in /37/ and /38/. A comparison between neural nets and the polynomial approach concerning relative effectiveness and discussing relative advantages and/or disadvantages is not known to me.

However, table 11 tries to give a relative evaluation of the properties of the different general interpolators discussed, which may help to select the best solution for a considered specific problem. In my opinion table 11 points out, that in the general case AMS is the best choice, although it lacks the feature of variable generalization.

	AMS	MIAS	Polynomial Approximation	Neural Net with Backpropagation
trainings-indication	yes	yes	no	no
response time	~ const.	$O(t)$ [1]	$O\left[\frac{(n+q)!}{n!\,q!}\right]$ [2]	$O(r)$ [3]
calculation effort	low	medium	high	medium
noise filtering	possible	difficult	possible	possible
tolerance against partial destruction	high	high	low	high
variable generalization	no	yes	no	no
convergence velocity	high	high	medium	low

Table 11 - (due to W. Mischo, M. Hormel) Qualitative evaluation of different interpolating memory approaches; [1] ν is here the number of trained points; [2] n is the number of inputs, q the highest polynomial degree (cf equation (41)); [3] r is the number of neurons used in the neural net.

II.8. Recapitulation

The chapter has dealt with locally generalizing or - what is the same - automatically interpolating memory systems. Starting from an eventual model for the human cerebellum an appropriate organisation for such a memory system was put forward and a special, very efficient implementa-tion called AMS was explained in detail. Its ability to store information therein was studied by developing test criteria and looking with their help at the properties shown in connection with the storage of some three-dimensional test functions. The inherent ability of AMS to reduce noise in the storage process was pointed out. Some theoretical considerations treated general storage capacity and convergence to the required memory response. As an extension the idea of variable generalization was introduced and a way to handle it with AMS. Furtheron a mathematically motivated alternative memory system concept, called MIAS was treated in this connection. By comparing AMS and MIAS, advantages and disadvantages of both concepts were discussed. Some direct application areas for locally generalizing and/or automatically interpolating memories were described. Finally, a short comparison of the here advocated approaches with the very often in a similar context favoured general neural net approach was given.

II.9. Literature

/1/ Albus, J. S.: Theoretical and Experimental Aspects of a Cerebellar Model. PhD Thesis, Univ. of Maryland 1972

/2/ Albus, J. S.: A New Approach to Manipulator Control: The Cerebellar Model Articulation Controller (CMAC) Trans. ASME, series G 1975

/3/ Pellionisz, A. J.: Robotic connected to neurobiology by tensor theory of brain function. IEEE Int. Conf. on Systems, Man and Cybernetics 1985

/4/ Albus, J. S.: A Model of the Brain for Robot Control - Part. 2: A Neurological Model. Byte 1979

/5/ Eccles, J. C.; Ito, M.; Szentagothai, J: The Cerebellum as a Neuronal Machine. Springer Verlag 1967

/6/ Albus, J. S.: Mechanisms of Planning and Problem Solving in the Brain. Mathematical Biosciences 45 1979

/7/ Ersü, E.; Tolle, H.: A New Concept for Learning Control Inspired by Brain Theory. Proceed. 9th IFAC World Congress Budapest 1984

/8/ Jansen, J. K. S.; Nicolaysen, K.; Rudjord, T.: Discharge Pattern of Neurons of the Dorsal Spinocerebellar Tract Activated by Static Extension of Primary Endings of Muscle Spindles. J. Neurophysiol. 29 1966

/9/ Kohonen, T.: Content-Addressable Memories. Springer Series in Information Sciences - Springer Verlag 1980

/10/ Ersü, E.; Militzer, J.: Implementation of a Neuron-like Associative Memory System for Control Applications. Proceed. of the ISMM 8th Int. Symp. MIMI, Davos 1982

/11/ Militzer, J.: Prüfung der Eignung von assoziationsfähigen Modellen der nervösen Informationsverarbeitung zur lernenden Regelung technischer Prozesse. DFG-Forschungsvorhaben To 75/9-4. 3. Zwischenbericht 1.1.84-31.12.84

/12/ Schwefel, H. P.: Numerische Optimierung von Computer-Modellen mittels der Evolutionsstrategie. Birkhäuser Verlag, Basel 1977

/13/ Strang, G.; Fix, G. J.: An Analysis of the Finite Element Method. Prentice Hall, Inc. 1973

/14/ Parks, P. C.; Militzer, J.: Convergence properties of associative memory storage for learning control systems. Avtomatika i Telemekhanika 5O 158-184 (in Russian), Automation and Remote Control 5O Pr. II 254-286, (in English). 1989

/15/ Ellison, D. On the convergence of the Albus perceptron. IMA Journal of Mathematical Control of Information 5, 315-331. 1988

/16/ Parks, P. C.; Militzer, J.: Improved allocation of weights for assosiative memory storage in learning control systems. Preprints of the First IFAC Symposium on "Design Methods of Control Systems", Zürich, Sept. 4-6, 1990.

/17/ Parks, P. C.; Militzer, J.: Convergence properties of associative memory storage for learning control systems. Proceed. IFAC Symp. on "Adaptive Systems in Control and Signal Processing", Glasgow, April 19-21, 1989 Vol. 2, 565-572

/18/ Barnett, S.; Matrices in Control Theory. Publishing Co. Malaba, Florida, 2nd edition 1984

/19/ Karczmarz, S.: Angenäherte Auflösung von Systemen linearer Gleichungen. Bull. Intern. Acad. Pol. Sci. Lett., Cl. Sci. Math. Natur. 1937

/20/ Aved'yan. E. D.: Modified Kaczmarz Algorithms for Estimating the Parameters of Linear Plants, Automatika y Telemekkanika 5, 1978

/21/ Aved'yan, E. D.; Hormel, M.: (submitted to Avtomatika i Telemekhanika)(in Russian) 1991

/22/ Militzer, J.; Tolle, H.: Vertiefungen zu einem Teilbereiche der menschlichen Intelligenz imitierenden Regelungsansatz. Tagungsband - Jahrestagung der Deutschen Gesellschaft für Luft- und Raumfahrt, München 1986

/23/ Ersü, E.; Tolle, H.: Acceleration of Learning by Variable Generalization for On-Line Self-Organizing Control. Vth Polish-English Seminar "Real-Time Process Control", Radzcejowice, Poland 1986

/24/ Ersü, E.; Tolle, H.: Vertiefende Untersuchungen zu gemäß der menschlichen nervösen Informationsverarbeitung lernenden Systemen. Abschlußbericht für das von der VW-Stiftung geförderte Vorhaben I/60/90/ TH Darmstadt 1988

/25/ McLain, D. H.: Drawing Contours from Arbitrary Data Points. Comp. J. Vol 17, 1974

/26/ Renka, R. J.; Cline, A. K.: A Triangle-Based C^1 Interpolation Method. Mount. J. of Math. Vol 4 1984

/27/ Tolle, H.; Militzer, J.; Ersü, E.: Zur Leistungsfähigkeit lokal verallgemeinernder assoziativer Speicher und ihren Einsatzmöglichkeiten in lernenden Regelungen. Messen-Steuern-Regeln (msr) 1988

/28/ Oh, S. Y.: A Walsh-Hadamard Based Distributed Storage Device for the Associative Search of Information. IEEE Trans. on Pattern Analysis and Machine Intelligence (PAMI) Vol. 6. 1984

/29/ Shepanski, J. F.; Macy, S. A.: Teaching Artificial Neural Systems to Drive: Manual Training Techniques for Autonomous Systems. 1st Int. Conf. on Neural Networks, San Diego 1987

/30/ Lippmann, R. P.: An Introduction to Computing with Neural Nets. IEEE ASSP Magazine, April 1987

/31/ Kolmogorov, A. N.: On the Representation of Continuous Functions of Many Variables by Superposition of Continuous Functions of one Variable and Addition. Dokl. Akad. Nauk. USSR 114 1957

/32/ Lorentz, G. G.: The 13th Problem of Hilbert. Proceed. of Symposia in Pure Math. 28. American Mathematical Society 1976

/33/ Hecht-Nielsen, R.: Kolmogorovs Mapping Neural Network Existence Theorem. Proceed. IEEE Int. Conf. on Neural Networks 1987

/34/ Moody, J.; Darken, Ch.: Learning with Localized Receptive Fields, Proceedings of the 1988 Connectionist Models Summer School, M. Kaufmann Publishers, Inc. 1989

/35/ Huang, W. Y.; Lippmann, R. P.: Neural Net and Traditional Classifiers - in Anderson, ed. "Neural Information Processing Systems". American Institute of Physics 1988

/36/ Kohonen, T.; Masisara, K.; Saramaki, T.: Phonotopic Maps - Insightful Representation of Phonological Features for Speech Representation. Proceed. IEEE 7th Int. Conf. on Pattern Recognition. 1984

/37/ Ivakhnenko, A. G. : Heuristic Self-Organization in Problems of Engineering Cybernetics. Automatica 6, 1970

/38/ Kortmann, M.; Unbehauen, H.: Ein neuer Algorithmus zur automatischen Selektion der optimalen Modellstruktur bei der Identifikation nichtlinearer Systeme. Automatisierungstechnik, (at), 1987

III. Macrointelligence

III.1. Learning control loop with an explicit process model - LERNAS

We define "macrointelligence" as psychologically-motivated action-oriented problem solving feedback structures in the brain. In this context the learning control loop with an explicit process model was introduced as a basic architecture. It was named "LERNAS" by E. Ersü and has been already described qualitatively in section I.5 and its layout is drawn in fig. 5 of chapter I. This general outline will not be repeated here, and we shall turn now to a more detailed mathematical presentation.

III.1.1. Assumptions and definitions

Throughout this chapter III we shall use the following assumptions and notation:

1. The learning control loop operates in a discretised fashion in time with the constant sampling time T and with the sampling taking place at times kT, $(k+1)T$, ... Normally kT is characterized by k only.

2. The process to be controlled is a deterministic, time-invariant or "weakly time-varying", single input/single output or multi input/single and/or multi output plant with unknown linear or non-linear dynamics. "Weekly time-varying" means in this context that process changes are much slower than changes in the process model generated by learning. The process states (describing the process when it is reduced to coupled first order dynamics) need not to be measurable.

3. The inputs $\underline{u}$ from the n_u-dimensional space $\mathbb{R}^{n_u}$ are constrained by upper and lower limits $\underline{u}_{max}$ and/or $\underline{u}_{min}$ and furthermore quantized, leading to a finite number of input values at each sampling instant. $\underline{u}(k)$ is applied at the plant input during the time-interval $(k+\delta)T \leq t < (k+1+\delta)T$ with $0 < \delta << 1$.

4. The measured values $\underline{y}$ from the n_y-dimensional space $\mathbb{R}^{n_y}$ - being usually the process outputs to be controlled - are also constrained by some $\underline{y}_{min}$, $\underline{y}_{max}$. $\underline{y}(k)$ is the value measured at time instant kT.

5. $\underline{y}$ and $\underline{u}$ are selected in a way such that all important behavioural features of the process can be observed and influenced.

6. Essential disturbances $\underline{v} \in \mathbb{R}^{n_v}$ are measurable and limited. $\underline{v}(k)$ is the value of $\underline{v}$ at $t = kT$.

7. The measurable process behaviour can be described sufficiently exactly by:

(1) $\quad\quad \underline{y}(k) = \underline{f}_p \left[\underline{\psi}(k\text{-}1), \underline{u}'(k\text{-}1), \underline{v}'(k\text{-}1) \right]$

Here $\underline{\psi}(k\text{-}1)$ represents a description of the status of the process at time instant $t = (k\text{-}1)\,T$ on the basis of a certain history of the measurable quantities (' = transposition):

(2) $\quad\quad \underline{\psi}(k\text{-}1) = [\underline{y}'(k\text{-}1), \underline{y}'(k\text{-}2) \dots \underline{y}'(k\text{-}\alpha),$
$\quad\quad\quad\quad\quad\quad\quad \underline{u}'(k\text{-}2) \dots \underline{u}'(k\text{-}\beta),$
$\quad\quad\quad\quad\quad\quad\quad \underline{v}'(k\text{-}2) \dots \underline{v}'(k\text{-}\gamma)]$

(α, β, γ are indicating the possibly different amount of history considered).

8. The goal of the control effort can be characterized by a performance criterion, which can be split into subgoals connected to sampling instants, so that with the demanded values $\underline{w}$ for $\underline{y}$ or a subset of $\underline{y}$ ($\longrightarrow \underline{w} \in \mathbb{R}^{n_w}$, $n_w \leq n_y$) one obtains as a general form for the performance criteria to be considered:

(3) $\quad\quad J(k) = L\left[\underline{y}(k+1), \underline{w}(k+1), \underline{u}(k)\right] \longrightarrow \text{Min } \underline{u}(k)$

$J(k)$ is structured in such a way that by selecting $\underline{u}(k)$ to minimise it the difference at time $t = (k+1)\,T$ between the demanded values and the outputs to be controlled will be diminished in the sense of some metric. Actually for $n_y > n_w$ not all elements of $\underline{y}$ will necessarily appear in $J(k)$ and possibly further quantities may be taken into account in $J(k)$, e.g. $\underline{u}(k\text{-}1)$, since a limitation of changes of $\underline{u}$ is normally of importance for the control quality from a practical point of view.

III.1.2. Functional description

We shall now discuss what happens in LERNAS at a certain sampling instant kT. For this purpose, three different tasks will be distinguished:

- sampling of process quantities and adaptation of the predictive process model;

- selection of an advantageous plant input by optimization;

- adaptation of the associative memory used in the controller and computation of the process input to be applied.

Furthermore for the sake of clarity in considering the detailed figures giving the required information about the functioning of the general structure from fig. I.5 we shall

- split up the block "assessment/optimization" into two elements "performance criterion" and "optimization";

- split up the block "control strategy" into two elements "optimal control " and "active learning"

- always show the elements not involved in the considered task in addition to the elements which are involved.

III.1.2.1. Sampling of process quantities and adaptation of the predictive process model (fig. 1)

At kT the measurable quantities $\underline{y}(t)$ and $\underline{v}(t)$ will be sampled and memorized in hold files as well as $\underline{u}(k-1)$ which is still active at this time instant (compare III.1.1.-3.); $\underline{\psi}(k-1)$ is changed into $\underline{\psi}(k)$. The predictive process model learns through the respective training process the mapping:

(4) $M_p: \left[\ \underline{\psi}(k-1),\ \underline{u}(k-1),\ \underline{v}(k-1)\right] \longrightarrow \underline{y}(k)$.

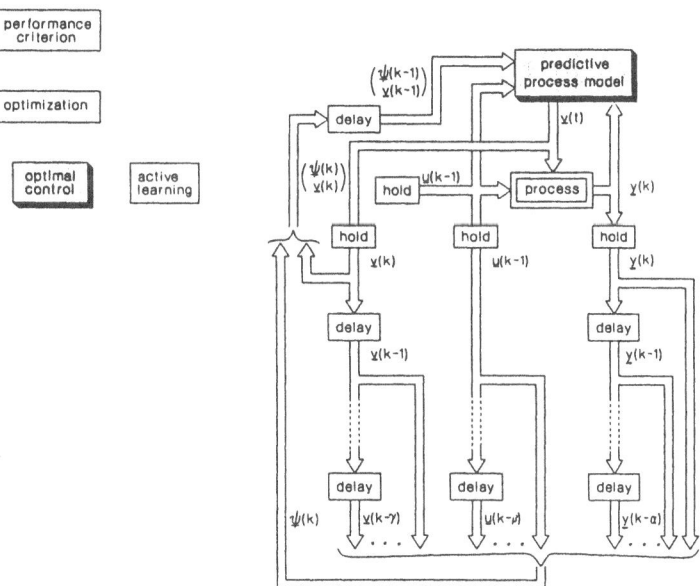

Fig. 1: Signal flow diagram comprising sampling and predictive model adaptation; non-involved blocks not connected, shaded blocks indicating associative memory systems (adapted from /1/).

Under the assumption from section III.1.1, that the process behaviour has only a weak direct dependency on the sampling value k (time t), the model is - concurrently with the functioning of the feedback control - stepwise adapted to generate an approximate reproduction of the functional process description f_p - see (1) -. The local generalization, which is a property of the associative memory system used for process model storage, ensures in this context that the adaptation is not restricted to the training points just encountered. In contrast to the actual process, where - short transition phases neglected - inputs and outputs have to be attributed to the same (actual) sampling instant kT, the connections learned by the predictive process model are relationships between situation describing values at kT and output values at the next sampling instant (k+1)T. This justifies the adjective "predictive" used to describe the process model.

III.1.2.2. Selection of an advantageous plant input by optimization (fig. 2.)

The long range goal of the optimization procedure is to build up the best possible control strategy for the process considered in all situations to be expected, making the learning level superfluous later on, or, in other words, to generate a mapping of the following kind:

$$(5) \qquad M_c^{-1}: [\underline{y}_d(k), \underline{\psi}(k), \underline{v}(k)] \longrightarrow \underline{u}(k) \approx \underline{u}(k)_{opt}.$$

In the beginning the respective associative memory system will be empty and/or trained by some random or heuristically estimated values, giving responses which may not be very useful or even misleading. However, the $\underline{u}(k)$ supplied for the actual control difference

$$(6) \qquad \underline{y}_d(k) = \underline{w}(k) - \underline{y}(k),$$

the situation $\underline{\psi}(k)$ and the disturbance $\underline{v}(k)$ will become better and better, as the process model, from which the best control input is derived, improves with time and when the number of encountered situations is dense enough to allow good interpolation to take place.

Since we are using a process model learned and/or built from numerically data we can also apply numerical procedures for the optimization. For such numerical algorithms some initial value has to be supplied. A natural choice for such a starting value $\underline{u}^{(0)}(k)$, is the actual response from (5), with the above mentioned property of being poor in the beginning, but becoming a better guess for the optimal value later on.

---

[1]It is not necessary that the amount of history included in $\underline{\psi}(k)$ is the same for the process model and the controller. Therefore in general a distinction has to be made between $\underline{\psi}_p$ in (4) with α_p, β_p, γ_p according to (2) and $\underline{\psi}_c$ in (5) with α_c, β_c, γ_c. This distinction has been dropped here for the sake of simplicity.

The optimization algorithms to be discussed in detail later on, generate from $\underline{u}^{(0)}(k)$ further eventual inputs $\underline{u}^{(1)}(k)$, $\underline{u}^{(2)}(k)$ ... $\underline{u}^{(x)}(k)$, for which the resulting process behaviour is estimated from (4) (i = 0,1 2...x):

(7) $[\underline{\psi}(k), \underline{u}^i(k), \underline{y}(k)] \rightarrow \hat{\underline{y}}^{(i)}(k+1/k).$

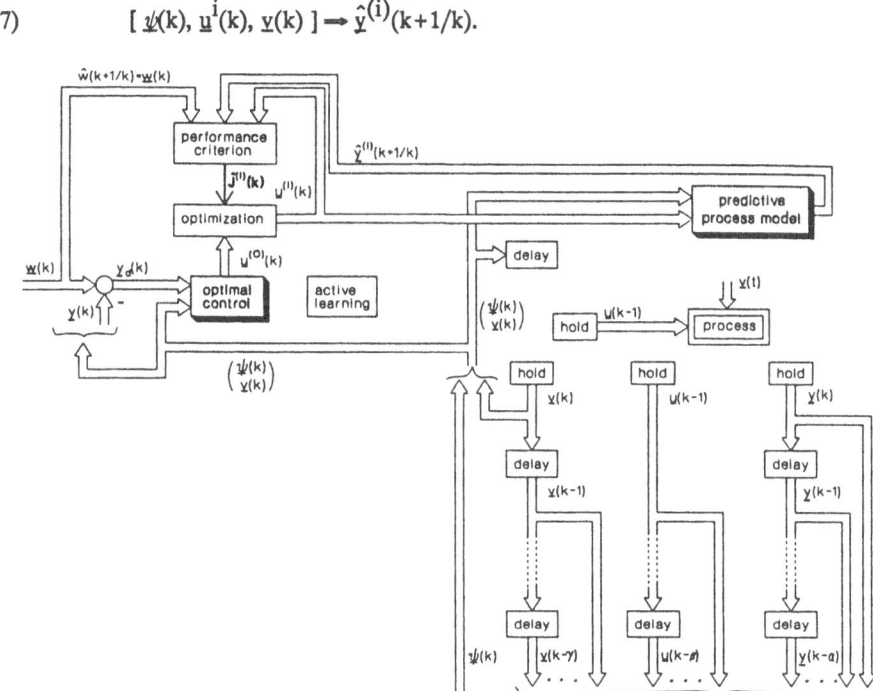

Fig. 2: (adapted from /1/) Signal flow diagram for the optimization; non-involved blocks not connected, shaded blocks indicating associative memory systems. $\underline{y}(k)$ out of $\underline{\psi}(k)$ is used for generating $\underline{y}_d(k)$ = $\underline{w}(k)$ - $\underline{y}(k)$. The connection $\underline{w}(k)$ - control strategy = from fig. I.5 is not appearing here, since $(\underline{y}_d, \underline{y})$ => $\underline{w}(k)$. (In fig. I.5 - the situation describing "past information" and $\underline{y}(k)$ were separated for the sake of clarity).

Feeding $\hat{\underline{y}}^{(i)}(k+1/k)$ together with some estimate $\hat{\underline{w}}(k+1/k)$ of $\underline{w}$ at time $(k+1)T$ - in general $\hat{\underline{w}}(k+1/k)$ = $\underline{w}(k)$ - into the performance criterion (2) gives performance estimates

(8) $\hat{J}^{(i)}(k) = L [\hat{\underline{y}}^{(i)}(k+1/k), \hat{\underline{w}}(k+1/k), \underline{u}^{(i)}(k)]$

which in turn can be used for further planning to reach the optimal $\underline{u}(k)$. To obtain meaningful results all optimization steps have to be performed, naturally, under the condition of sufficiently reliable responses from the process model monitored by some training indication $\eta \geq \eta_{min}$ as discussed in section II.2.4.

The search with respect to the best $\underline{u}^{(i)}$ is terminated by reaching either some lower bound in the change of the performance criterion or the boundary of the already explored area of process behaviour or, possibly, some limit set for the time span allowed for the search procedure. The $\underline{u}^{(\sigma)}$ then reached is considered as the actual estimate $\hat{\underline{u}}^{*}(k)$ of an advantageous control input for the examined situation.

III.1.2.3. Adaptation of the associative memory used in the controller and computation of the process inputs to be applied (fig. 3).

The control vector $\hat{\underline{u}}^{*}(k)$ determined by the optimization procedure is stored together with the actual control situation as the mapping (4):

(9) $[\underline{y}_d(k), \underline{\psi}(k), \underline{v}(k)] \longrightarrow \hat{\underline{u}}^{*}(k)$.

With continuing process control a control strategy covering all process situations is built up stepwise and gradually improved towards optimality in the sense of the performance criterion adopted.

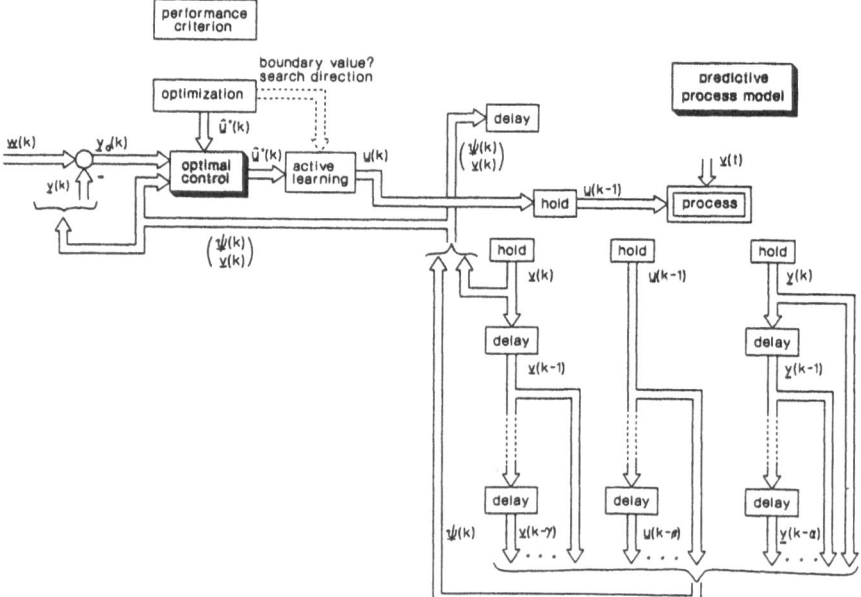

Fig. 3: Signal flow diagram for controller adaptation and plant input generation; non-involved blocks not connected, shaded blocks indicating associative memory systems (adapted from /1/). For details concerning the active learning compare the text.

However, this is only the case if one has installed some control loop starting philosophy, e.g. to use at first - when the process model memory is still empty so that $\eta \geq \eta_{min}$ is nowhere achieved - some estimated control input to obtain the desired process response for a considered set point. Furthermore, the process model is expanding only very slowly, the expansion being solely due to the local generalisation effect together with $\eta_{min} < 100$ %. On the other hand one would expect that if one has obtained, by the optimization procedure, a path to the boundary of the already explored process behaviour area with a non-zero slope of the performance criterion reduction, then a step across the boundary in the direction by which this boundary was reached, will bring a further decrease of the performance criterion. As long as this step is not too big, it can be assumed that by such an explorative action one improves not only the process reaction but also the knowledge about the process behaviour. Now, the active learning block from fig. 1-3 comprises just the initial control value given from outside and the just stated procedure of exploration into unknown process behaviour with a free parameter for fixing the size of the exploratory step into the area of still unknown process behaviour. The details will be explained together with the optimization procedure, since they depend in detail on the respective numerical algorithms.

III.1.2.4. ν steps look-ahead capability

By planning a series of process inputs $\underline{u}(k)$, $\underline{u}(k+1)$, ... $\underline{u}(k+\nu-1)$ instead of a single control action and by recursive use of the process model to estimate the process reactions $\hat{\underline{y}}(k+1)$, $\hat{\underline{y}}(k+2)$ ... $\hat{\underline{y}}(k+\nu)$ one may use a performance criterion, for which the evaluation interval stretches over several sampling steps. The controller looks then further ahead, giving in general an improved closed loop behaviour for the overall control problem. However, this means an additional computation load and is not achievable in the beginning when the process knowledge is still poor. One possibility is, to work at first with a one step look-ahead criterion and to turn later on to a several steps look-ahead criterion for improvement of the controller.

III.1.3. Optimization

An important feature of the considered learning control loop LERNAS is the self-organization of the controller by optimization. Actually, an optimization problem is characterized by the nature of the performance criterion, stating the goal of the optimization, and the constraints, such as the differential equations describing the dynamic behaviour of the process and limitations such as maximal values allowed for inputs and possibly process states. Such problems are in themselves not easy to solve, and a great variety of appropriate approaches have been explored without uncovering a best or even an entirely satisfactory way to handle them (see e.g. /2/). In our case the situation is even more complex: we have a largely incomplete description of the process, which is updated and improved, or at least partly changed, in a step by step procedure. This leads to the necessity of considering in practice subgoals instead of the overall goal. Taking these facts together, we consider continuously changing optimization problems, which make a rigorous mathematical treatment impossible.

However, we know, that in general some not necessarily unique solution to our problem exists and we are further more satisfied with a fair approximation to any of the eventually numerous possible solutions. To clarify the existence of a solution let us consider the task of eliminating the difference between the actual plant output and the desired output value. In the case that this is possible at all - which, for example, means that we are not requiring a room to reach a temperature higher than the maximum temperature of the heating device - we can imagine building up as a controller the inverse of the process together with an integrator, so that the whole open loop transfer function of the control loop is fixed as $1/s$ and consequently the closed loop transfer function is $1/1+s$, which leads to the desired total cancellation of the control difference as $t \Rightarrow \infty$. So in principle an appropriate controller exists, if we allow for rich enough dynamics, which in our case is equivalent to a rich enough history in $\underline{\psi}_c(k)$.

With the picture of subgoals and evolving process knowledge in mind it is clear that general statements on convergence and the best ways to proceed are not possible, and we shall turn instead to a discussion of the approaches actually used and experiences gained. The results will be stated only qualitatively here as supporting quantitative simulation values will be presented later on.

Regarding performance criteria, it is obvious, that they always contain the control difference. The reduction of this value is not only an important subgoal but also a general overall control goal. But the usual preference to use its squared value in the performance criterion is not necessary, since this preference is due to the easy analytical treatment of squared values. However, in our case only purely numerical optimisation procedures are possible and therefore the use of the absolute value of the control difference presents no additional difficulty and has also been sometimes used with success in the simulations.

A further point to be noted is that one cannot expect the optimization results to contain aspects which have not been taken into account explicitly in the performance criterion. So a performance criterion comprising the control difference only may lead to relatively frequent control actions, an unwanted behaviour, which can be easily avoided by including a control energy measure into the criterion.

Turning now to optimization procedures it should be clear that simple direct optimization by searching for points giving performance criterion reductions is the most suitable procedure. Through the quantization of process inputs and outputs and their representation by whole numbers (compare section II.2.2) we are dealing with the optimization of the input value $\underline{u}(k)$ using a search procedure in a n_u-dimensional grid (n_u being the dimension of the vector $\underline{u}$), in which the computation of derivatives and/or gradients is not too helpful.

Therefore, the first choice was a Hooke-Jeeves-algorithm (/3/, /4/). This algorithm alternates between search cycles and extrapolation actions. Searching takes place stepwise in all coordinate directions, at first in a positive, then in a negative direction. If a search step does not lead to a trained situation and/or in a trained situation to a lower value of the performance index it is deleted. Otherwise the new point is used as basis for the next steps in the search.

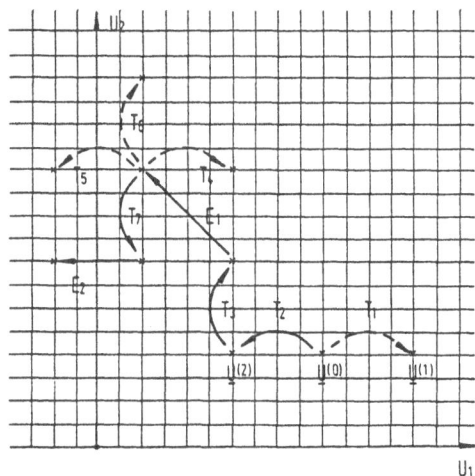

Fig. 4: Twodimensional example of first steps of the used Hooke-Jeeves-algorithms: $h_0 = 4\epsilon$; * points where the performance criterion is evaluated; - - -> search step T without, ——> search step T with success => successful extrapolation E (followed by search steps again).

Extrapolation is used to try to accelerate searching by immediately going further in promising directions. In the beginning, the first two successful search steps are used for generating the extrapolation direction (see fig. 4). Although the step size in the extrapolation direction may be set up freely, it seems prudent to couple it to the step size h actually used. In our case we just continue to the next grid point of a grid with mesh-width h. From this new point a new search cycle is initiated without considering the accompanying performance criterion value. If in this search a point is found with a lower performance criterion value than the performance criterion value at the starting point of the extrapolation, the extrapolation is classified as being successful. A new extrapolation is then made from this new point in the same direction as the preceding successful extrapolation. This is continued until an unsuccessful extrapolation is reached. Such an unsuccessful extrapolation is deleted like an unsuccessful search, and the search cycle is continued from the starting point of the unsuccessful search. If then no successful extrapolation is found, the step size h is divided by two and the search startet anew. The algorithm needs only two parameters, the starting step size h_0 and some minimal step size, which have to be set by the user.

With respect to active learning, since only grid points should be considered, one projects at first the eventual active learning direction onto the coordinate directions and searches along these with the step size b_0 in some cycle, e.g. in the order of increasing indices. In certain cases this may not lead to a success, e.g. when the active learning direction lies in a coordinate direction and this coordinate direction is parallel to the boundary of the known territory. Consequently one has to include in the optimization procedure in case of failure of these first steps, the fact that the other coordinate-system directions have also be searched.

The Hooke-Jeeves algorithm is not the only one designed for direct optimization without the use of derivatives/gradients. Very interesting alternatives are the evolution strategies, which try to imitate biological optimization principles of selecting for survival the best adopted species from a variety of species generated by mutation. An extensive review can be found e.g. in /5/. Since the learning control loop is also inspired by a biological background, it was natural to try evolution strategies for the optimization problem too. Simulation results indicated that the simplest of these strategies seems to be also the most effective one in LERNAS, although it has to be admitted , that not all possible parameter variations have been tried out for the more complex strategies, due to the large number of free parameters existing in these strategies.

The simplest evolution strategy comprises the following three different steps:

- mutation: from a parental individual characterized by n "genes" a descendant is generated by a slight change of the genes.

- vitality evaluation: to the genes some "vitality" is attributed and for both individuals, the parental individual and the descendant the level of vitality is computed.

- selection: that individual, which has the higher vitality is used as the new parental individual for the next mutation step.

With respect to our optimization problem we have to identify the "genes" with the components of the $\underline{u}$-vector (individual). Mutation means a random change of these components. "Vitality evaluation" means the computation of the performance criterion value and identification of a lower value with a higher vitality and - finally - selection means the use of a better point in $\underline{u}$-space for the next step. In detail we can write the algorithm as follows (minimization):

- step zero: select a starting point $\underline{u}$ and put $\underline{u}^{(g)} = \underline{u}$, $g = 0$, $i = 0$
                 compute the performance criterion $J(k)$ for $\underline{u}^{(g)}$ as $J^{(g)}$.

- step one: compute $\underline{z}^{(g)}$ and put $\underline{u}^{(g+1)} = \underline{u}^{(g)} + \underline{z}^{(g)}$

- step two: compute the performance value $J^{(g+1)}$

- step three: set $\underline{u}^{(g)} = \begin{cases} \underline{u}^{(g+1)} & j = 0 \quad \text{for } J^{(g+1)} < J^{(g)} \\ \underline{u}^{(g)} & j = j+1 \quad \text{otherwise} \end{cases}$

- step four: if $j = j_{max}$ stop
  otherwise put $g+1 \rightarrow g$.

For the random changes $\underline{z}$ of the components of $\underline{u}$ we require in accordance with the properties of the natural evolution:

- expectation (linear average) $E(z_i) = 0$.
- variance $\sigma_{z_i}^2$ small

This leads to a Gauss-distribution:

$$(10) \qquad f(z_i) = \frac{1}{\sqrt{2\pi}\ \sigma_{z_i}} \cdot e^{-\dfrac{z_i^2}{2\sigma_{z_i}^2}}$$

for the generation of the random vector $\underline{z}$.

The standard deviation σ_{z_i} is called, in analogy to other search strategies the step size, since it represents the mean value of the length of the random search steps. The counter j can be understood as a convergence criterion: if the number of unsuccessful changes to reduce the performance criterion value reaches the chosen limit j_{max}, one assumes that no further reduction of this value is possible, which means the minimal value has been reached. This may be - as with the Hooke-Jeeves-algorithms - the global minimum or a strong local minimum. However, due to the random search, it is less likely that the evolution strategy is trapped in a local minimum than for other search procedures such as the Hooke-Jeeves strategy.

With respect to active learning the same procedure can be used for the evolution strategy as was explained earlier: one goes preferably in the direction from the last $\underline{u}$-value before the final $\underline{u}$-value to the final $\underline{u}$-value in the sense that one uses at first the direction of that component of $\underline{u}$, which was maximally changed, then if no success is achieved by going in this direction by the amount of b_0, one chooses the direction of the next smaller change and so on until all changed components are exhausted. If still no untrained point is reached, one tries the remaining (unchanged) components - if there are any - in the order of their indices. No success at all means that the point with a minimal value of the performance index is well in the interior of already explored territory.

With this remark we will conclude this general description of the optimization approach. Further information on the respective behaviour and relative merits of differences in performance criteria and optimization algorithms will be given later on in connection with simulation results for selected examples.

III.1.4. Test processes

Learning control loops are most interesting with complex processes, which are either difficult to model so that one would like to save the effort required, or which are not yet fully understood, so that modelling seems to be very difficult or impossible.

However, to clarify the behaviour of such control loops and to obtain some understanding about the influence of those parameters which have to be selected (put equal to specific values) in case of any real application (as e.g. r^*, the amount of local generalization) tests have to be performed with simulated processes, defined by some mathematical equations, but known to the learning control system only by their input and output values. This section will present, therefore, two examples, which have been used for investigations of LERNAS in this context. Further test processes have been taken into account in addition but will not be discussed here, since they have not been studied to the same extent and also since they did not lead to significantly different results.

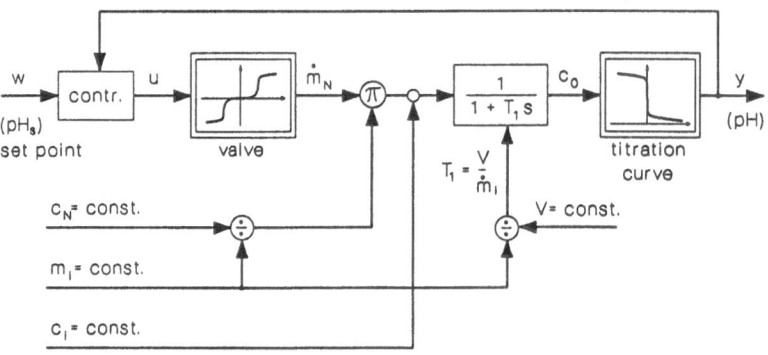

Fig. 5: Block diagram of pH neutralization process from fig. I.6.

The first test example is the pH-control introduced already in section I.4. to give some general picture of LERNAS behaviour. The technical layout of such a pH neutralization stirred tank reactor was sketched earlier in fig. I.6. The corresponding block diagram is given in fig. 5. It is already simplified in so far as a constant temperature level (25°C) was assumed. Since the process has been used solely as a test example, only the most difficult nonlinearity, the titration curve, was taken into account exactly in the computations. The nonlinear characteristic of the valve was substituted by a linear relationship with a limitation of the control action:

(11a) $0 \leq u \leq u_{max}$; $\dot{m}_N = u$

All other nonlinear effects are anyhow reduced to proportional elements by keeping the input - or as in the case of adaptation of the time constant T_1 both inputs - constant. Thereby one obtains the dynamics up to the input of the titration curve (compare fig. 5) as:

(11b) $\dot{c}_o = - \dfrac{1}{T_1} c_o + c_i + \dfrac{c_N}{\dot{m}_i} u$; $T_1 = \dfrac{V}{\dot{m}_i}$

The pH-value (output) and the titration curve can be described as in /6/ by the equations:

(12a) $pH = - \log [H^+]$

(12b) $[H^+] = \begin{cases} c_o/2 \left(\sqrt{1+4k_w/c_o^2} - 1 \right) & \text{for } c_o < 0 \text{ (alkaline)} \\ \sqrt{k_w} & \text{for } c_o = 0 \\ c_o/2 \left(\sqrt{1+4k_w/c_o^2} + 1 \right) & \text{for } c_o > 0 \text{ (acidic)} \end{cases}$

with k_w being the dissociation constant of water equal to

(13) $k_w = 10^{-14} \dfrac{mol^2}{l^2}$ (at temperature 25^0 Celsius).

Details of the performance criterion, the mappings and values used for mass flows $\dot{m}$, ionic concentrations c and so on will be given in connection with the specific simulations.

The second test example is a chemical reactor with two inputs, the massflow into the stirred tank reactor u_1 and the massflow through an additional heating device u_2, and two states - see fig. 6 - the concentration $A = x_1$ and the temperature $T = x_2$ in the reactor, which are assumed to be measured and by thus to represent also the outputs to be regulated y_1, y_2. With the following definitions:

R   gas constant

E   activation energy

(14) k_o frequency factor

$-\Delta H$ enthalpy of exothermic process

α overall heat transfer coefficient

F_H heat transfer area of the coil

c_p heat capacity of reactants

ρ mass density of reactants

c_{pH} heat capacity of heating fluid

ρ_H mass density of heating fluid

T_H temperature of heating fluid

we obtain due to /8/, from the mass balance and the temperature balance, the following differential equations for the values to be measured:

$$(15a) \qquad \dot{x}_1 = \frac{u_1}{V}(A_z - x_1) - x_1 \cdot k_o \cdot e^{-\frac{E}{Rx_2}} \qquad ; y_1 = x_1$$

$$(15b) \qquad \dot{x}_2 = \frac{u_1}{V}(T_z - x_2) + \frac{(-\Delta H)}{c_p \rho} x_1 k_o e^{-\frac{E}{Rx_2}} +$$

$$+ \frac{c_{pH}\rho_H \cdot u_2 (T_H - x_2)}{1/2 + \frac{c_{pH}\rho_H}{\alpha \cdot F_H} u_2} \qquad ; y_2 = x_2$$

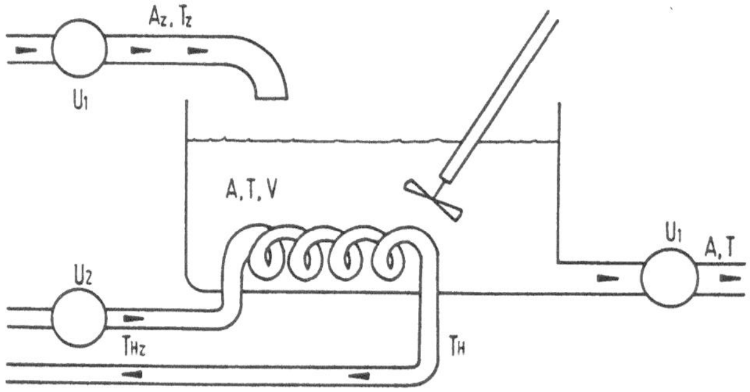

Fig. 6: Chemical stirred tank reactor with heating, $u_{1,2}$ representing the changable mass flows, A_z, A the concentrations in the input and in the reactor, T_z, T and $T_{H_z} \approx T_H$ the temperatures in the inflow and outflow of reactor and heating device, V the volume to be handled in the reactor (from /7/).

One sees immediately that there exists an exponential nonlinearity, which makes the process very difficult to control. This can be easily checked by setting up the equations on a computer and feeding in u_1, u_2 through some "knobs" to be handled by a test person. This person has a lot to learn, before he is able to avoid instability, which would mean in reality no output or an explosion.

Again numerical values of the coefficients in the equations will be given later on together with the specifications regarding performance criteria and mappings in connection with the discussion of simulations using LERNAS to control the process.

III.1.5. Results and experiences

III.1.5.1. LERNAS' handling of the test processes

The pH neutralization process in fig. 5 is an example of a highly nonlinear process with one input and one output. The following results cited here have been reported originally in /6/. The numerical values used in equation (11) were:

(16) $c_i = 10^{-3} \text{mol/l}$; $c_N = 0,1 \text{ mol/l}$; $\dot{m}_i = 50 \text{ l/sec}$; $V = 5000 \text{ l}$

A one step look-ahead performance criterion (c.f.(3)):

(17) $J(k) = [w(k+1) - \hat{y}(k+1)]^2 \longrightarrow \text{Min}$

was implemented, and no history was taken into account in the mappings (4) and (5) which leads to

(18) $\psi(k) = y(k).$

However, since later on the influence of changes in c_i, $\dot{m}_i$ were considered, these reactor inputs were treated as measurable disturbances so that one obtains for (4), (5) in addition:

(19) $\underline{v}'(k) = [e_i(k), \dot{m}_i(k)].$

The following quantizations were used in the predictive process model and control memory of AMS-type ($r^* = 16$, 12k Byte in both cases):

(20) $pH = y = 0,05$ $\dot{m}_N = u = 1 \text{ ml/sec.}$

$c_i = v_1 = 2 \cdot 10^{-7} \text{ mol/l}$ $\dot{m}_i = v_2 = 0,05 \text{ l/sec.}$

As the sampling time and the input limitations

(21) $T = 2 \text{ sec}$; $0 \leq u \leq 8 \text{ l/sec}$

have been chosen, and as the initial task conditions:

(22)   pH(0) = 3,75   ;   w = const = 9

empty memories (all values in the AMS-memory cells set to zero).

The learning behaviour has been displayed already in the figs. I.8 and I.9. After a first phase of trials the demanded value of w = 9 is reached in a fairly direct fashion even in the first run. The first repetition, the second run, shows furthermore an unexpectedly good behaviour, since the next runs improve the time to reach w = 9, but pay for that at first with some undesirable overshoot, until at the 68th run a fast change to pH = 9 is combined with practically no overshoot. Fig. 7 supplements fig. I.8, I.9 by giving the respective values of the sum of performance criterion values (17):

(23) $$J_\Sigma = \sum_{k=0}^{1000} J(k)$$

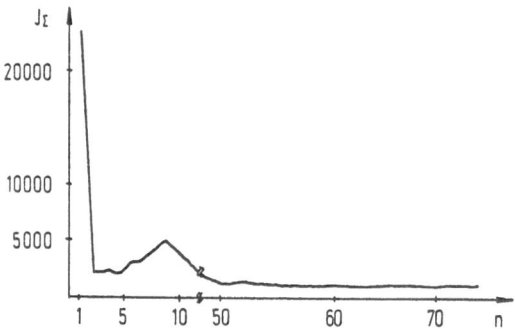

Fig. 7: Overall performance criterion value -(23)- in dependence of the number of runs from the same initial conditions to the same required pH-value w = 9 (For y = pH and u = $\dot{m}_N$ see fig. I.8/I.9). n = number of runs/transitions from initial value to w = 9.

As a learning curve fig. 7 appears at first surprising, since normally such curves show a monotonic decay. Here after the usual high minimization gain in the first steps we have an intermediate loss of performance before reaching a very low final value. The reason for this behaviour is active learning. By exploring the unknown process behaviour the performance can deteriorate in the intermediate phase before a very good performance is finally obtained.

Learning does not occur only by repetition, but also by task changes. At least the content of memory cells is changed by all new requirements. This can be seen from fig. 8 and 9. Here a sequence of required pH-values is asked for, ending with the most difficult situation of pH = 7 (cf.

to the titration curve I.7). One finds from fig. 8, that LERNAS is able to handle this demand up to the end of w = 7. From fig. 8 one sees, as well the much improved behaviour in the second run, the different behaviour in this run in reaching w = 9 compared with the same task without learning to move to w = 8, w = 7 (cf. fig. I.8).

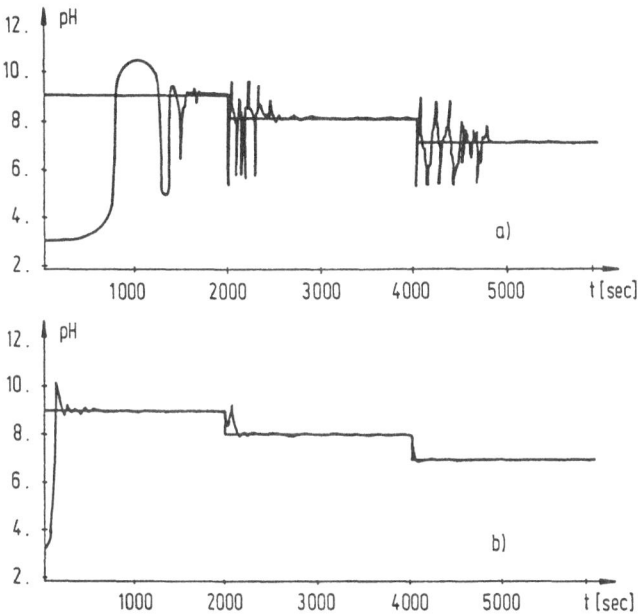

Fig. 8: First run - a)- and second run - b)- for a sequence of different demanded pH- values.

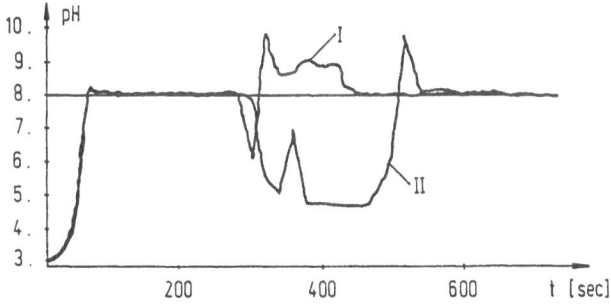

Fig. 9: Handling of impulsive disturbances by the learned nonlinear controller, learning loop of fig. I.5 disconnected. I impulsive disturbance of input mass flow $\dot{m}_i$ by 1 % at t = 280 sec, II impulsive disturbance of input ionic concentration of acid waste water c_i by 0,1 % at t = 280 sec.

Due to the local generalization, the learned controller (100 runs) is able to handle certain impulsive disturbances without further learning. Fig. 9 demonstrates this. Here the "learning loop" of fig. I.5 is disconnected and at t = 280 sec. an impulsive disturbance is added, in case of curve I of $\Delta \dot{m}_i = 0,5 \, 1/\text{sec}$, in case of curve II of $\Delta c_i = 10^{-6} \, \text{mol/l}$. The nonlinear trained controller can handle both cases (applied separately).

Also measurement noise can be handled by the trained nonlinear controller without the aid of the learning loop. Fig. 10 shows the controlled process behaviour (learning loop inactive) for an additive measurement noise on y with zero mean, uniform distribution and a range of [- 0,1; +0,1].

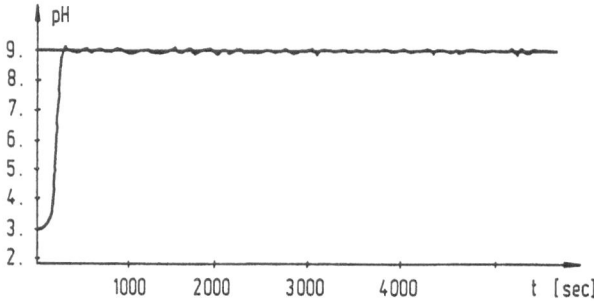

Fig. 10: Handling of measurement noise by the learned non-linear controller for w=8 - zero mean, uniform distribution, twice as big as the process output quantization - (learning loop disconnected)

Difficulties arise with continuous disturbances, since this means actually a change of the process to be controlled for which the controller was not trained. Fig. 11 displays the behaviour with the same disturbances as used for fig. 9, but now with constant changes of $\dot{m}_i$, c_i. Without the aid of the learning loop the controller cannot reach the demanded value (trained value) of w = 8, however after activation of the learning loop this is no longer a problem.

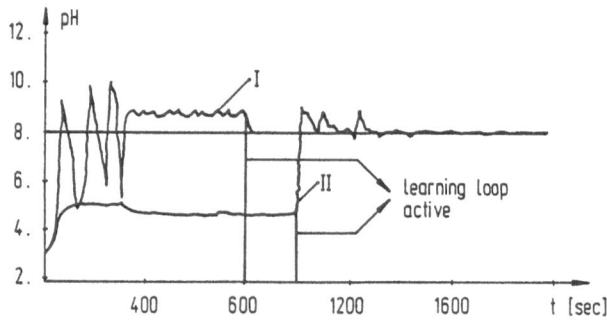

Fig. 11: Handling of constant changes of $\dot{m}_i$, c_i by 1 % and/or 0,1 % by a controller trained for w = 8 and the undisturbed $\dot{m}_i$, c_i - values of (16), at first with the learning loop disconnected, then with the learning loop activated. Curve I: $\dot{m}_i = 50,5 \, 1/\text{sec}$, curve II: $c_i = 1,001 \cdot 10^{-3} \, \text{mol/l}$.

The applicability of LERNAS to multivariable highly nonlinear processes has been investigated by using the stirred tank reactor of fig. 6. During the large number of simulation runs some different numerical values have been considered. However since no great change in behaviour occurred in consequence of this, only one set of data, which was used in the major part of the simulations, will be listed in the following:

Using appropriately scaled numerical values, the governing differential equations (15) can be written

$$(24) \qquad \dot{x}_1 = \frac{5 - x_1}{12,5} u_1 - 18,828 \cdot 10^{33} x_1 \cdot e^{-75,2315/x_2}$$

$$\dot{x}_2 = \frac{1}{400} \left[(24 - 32 \cdot x_2) u_1 + 242,88 \cdot 10^{33} x_1 \cdot e^{-75,2315/x_2} + 28,8 \frac{3,73 - 4x_2}{5 + 0,92u_2} u_2 \right]$$

$$y_1 = x_1$$
$$y_2 = x_2$$

As respective ranges and initial conditions one may use:

		min. value	max. value	initial value
concentration	x_1	0,0000	3,0000	0,0000
temperature	x_2	0,6825	0,9475	0,7330
reaction fluid massflow	u_1	0,0000	0,5360	0,0000
heating fluid massflow	u_2	0,0000	15,0000	0,0000

(25)

In the one step look-ahead performance criterion

$$(26) \qquad J(k) = \{ [w_1(k+1) - \hat{y}_1(k+1)]^2 + 10 \, [w_2(k+1) - \hat{y}_2(k+1)]^2 \} \, T$$

it was taken into account that the temperature is less sensitive to requirement changes than the concentration; the sampling time was included to allow the possibility of comparing results for different sampling times easily. As desired output values mainly the status

$$(27) \qquad w_1 = y_1 = x_1 = 1,43 \quad ; \qquad w_2 = y_2 = x_2 = 0,9225$$

was considered.

As in the case of the pH process no history has been included in the locally generalizing memory inputs, so that one obtains in accordance with (18):

(28) $\underline{\psi}(k) = \underline{y}(k)$

but here also no vector $\underline{v}$ of measurable disturbances was considered, leading to

(29a) M_p: $[\underline{y}(k), \underline{u}(k)] \longrightarrow \underline{y}(k+1)$

(29b) M_c: $[\underline{y}_d(k), \underline{y}(k)] \longrightarrow \underline{u}(k)$

(cf. (4), (5)). The main parameter values for the control loop and/or the memories of AMS type used were:

(30)
sampling time	$T = 2$ sec
training indication	$\geq 80 \%$
quantization of inputs	$\epsilon = 1 \%$ in all cases
memory cells AMS (29a)	$r_p = 8192$
memory cells AMS (29b)	$r_c = 1024$
active cells (29a),(29b)	$r^* = 32$
Hooke-Jeeves default mesh size	$h_0 = 32$ (cf. fig. 4)
active learning step size	$b_0 = 8$ (cf. III.1.3).

The first task considered was, to reach the desired status (27) from the initial values given in (25) with empty AMS memories at the beginning. However, at the plant input those values $\underline{u}_{st}$ = const were also added, which allow one to hold the demanded $\underline{w}$, if it is reached, exactly. This additional constant input is usually sufficient to reach the required w_2-value and to hold it, but one does not obtain the concentration y_1 close to w_1 (see curves in fig. 12). After roughly 20 repetitions, LERNAS has learned to handle the stirred tank reactor (fig. 12), but it takes still 20 repetitions more to bring the summed performance criterion value

(31) $J_\Sigma = \sum\limits_{k=1}^{50} J(k)$

down to a monotonic low value - see fig. 13.

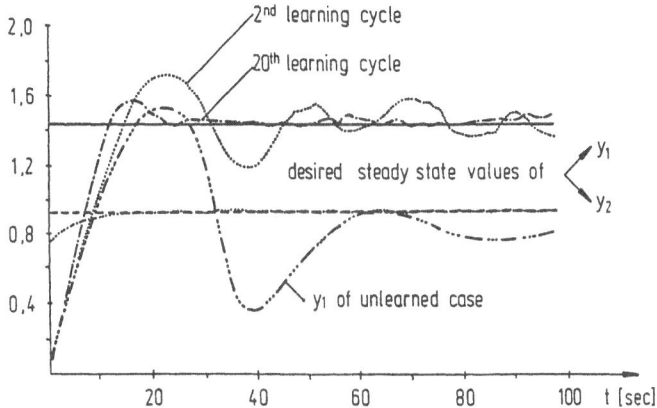

Fig. 12: Control of stirred tank reactor (fig. 6): unlearned case = constant input of $\underline{u}_{st}$, the required values to keep the reactor on the considered set point. 2nd, 20th learning cycles with LERNAS starting from empty memories and feeding learned $\underline{\Delta u}$- values ($\underline{u} = \underline{\Delta u} + \underline{u}_{st}$).

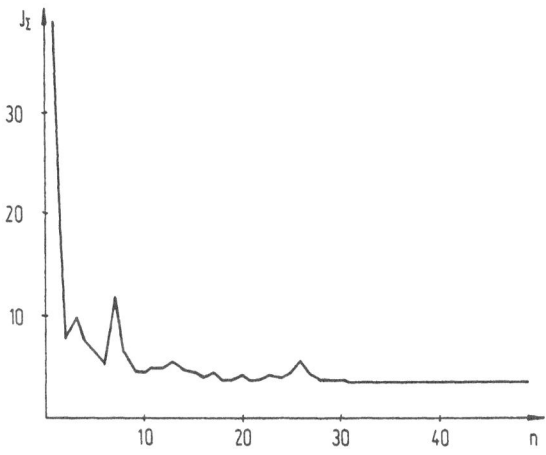

Fig. 13: Learning curve for the stirred tank reactor. Active learning prevents a further monotonic decay after the first big learning achievements until later on practically no unexplored process states of relevance for the control task exist.

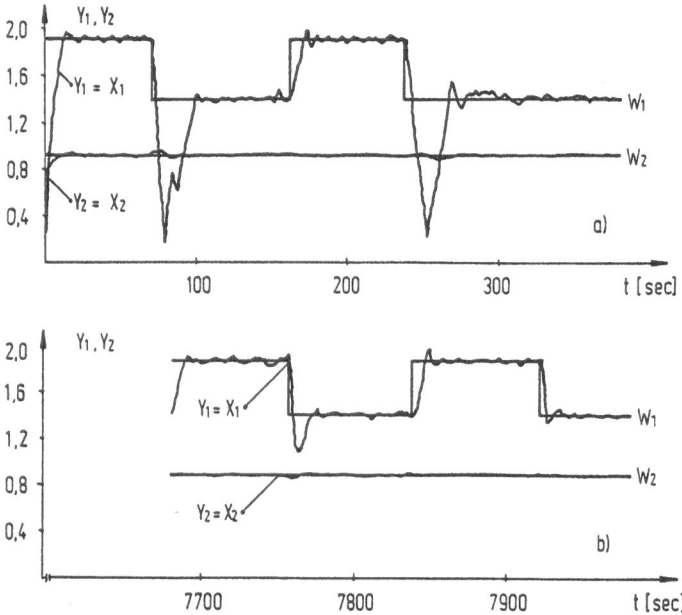

Fig. 14: Behaviour of stirred tank reactor with LERNAS if initially only the upward jumps from the initial conditions to the two different set points are trained. Since, during the first set point changes (a)) a high overshoot exists for downward jumps, this improves by additional learning and after 50 changes a fully satisfactory situation is reached (b)).

Fig. 14 shows, that for a strongly nonlinear multivariable process like the stirred tank reactor, a training routine to reach certain desired states $\underline{w}^I$ and $\underline{w}^{II}$ from the initial conditions $\underline{x}_0$ in a satisfactory way is not enough to guarantee also a satisfactory transition from $\underline{w}^I$ to $\underline{w}^{II}$, which goes in the opposite direction to the transitions from $\underline{x}_0$ to $\underline{w}^I$, $\underline{w}^{II}$. New process regions have to be explored to handle this situation. The situation displayed in fig. 14 is described by the rectangular set point course:

$$(32) \qquad \underline{w}(k) = \begin{cases} \underline{w}^I = \begin{bmatrix} 1,91 \\ 0,9200 \end{bmatrix} & \text{for} \quad n \cdot 160 \le t < n \cdot 160 + 80 \\[2ex] \underline{w}^{II} = \begin{bmatrix} 1,43 \\ 0,9225 \end{bmatrix} & \text{for} \quad 80 + n \cdot 160 \le t < n \cdot 160 \end{cases}$$

$$n = 0, 1, 2 \ldots 49$$

and fig. 14a shows the first four cycles, fig. 14 b the last four cycles: Since the jumps upward are all nicely in accordance with the original training of upward jumps, the downward jumps are, at the beginning, unacceptable from a control standpoint of view, but they improve to a fair standard in course of the repetitions.

Although in the case of the stirred tank reactor disturbances $\underline{v}$ were not included explicitly in the process model, one can adapt this case to investigate the consequences of unmeasurable and measurable disturbances. The basic trick is to control only one of the outputs - in fact the more critical one, the concentration - leaving the second one, the temperature, free. One input, e.g. the mass flow of the heating fluid, is then considered as a disturbance, since the other one is used as a control input. This idea was applied with a regular change of the mass flow of the heating fluid described by:

$$(33) \qquad u_2 \equiv v = \begin{cases} 6,10 \ \text{for} \quad n \cdot 160 \leq t < n \cdot 160 + 80 \\ 0,20 \ \text{for} \ 80 + n \cdot 160 \leq t < n \cdot 160 \end{cases}$$

$$n = 0,1,2 \ldots 49.$$

Fig. 15 shows the behaviour of the process being brought up at first to the desired set point $\underline{w}' = (1,43; \ 0 \ 9225)$. One sees that the oscillation of u_2 leads to big deviations of the concentration y_1 from its desired set point.

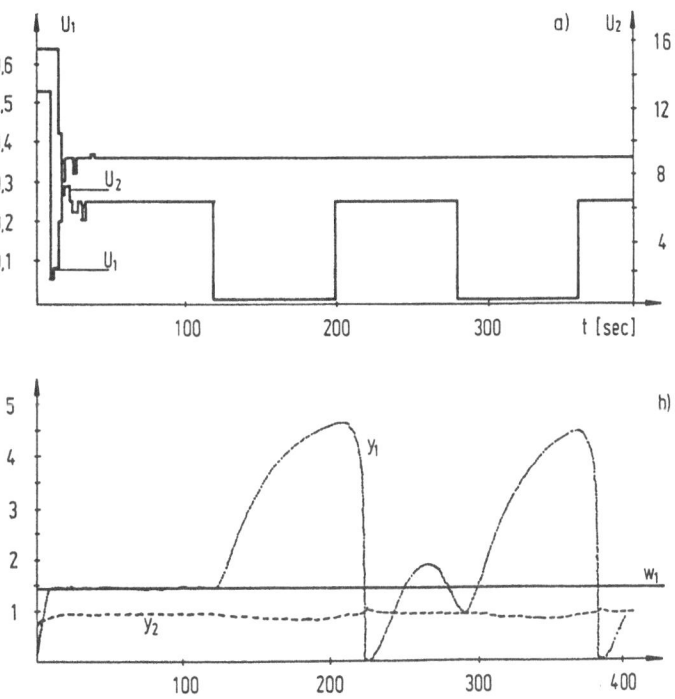

Fig. 15: Stirred tank reactor with exact u_{1st} to keep some achieved set point and periodic deviations of u_2 from the respective u_{2st} (a)). The concentration y_1 shows big deviations from its desired set point value w_1 due to the oscillations of u_2 (b)).

The control task for LERNAS is now set up in the following way: the AMS representing the predictive process model is trained to bring the process from the initial conditions to the nominal set point $\underline{w}' = (1{,}43\ ;\ 0{,}9225)$. A new performance criterion is defined:

(34) $J(k) = [w_1 - \hat{y}_1(k+1)]^2$

and the so far untrained controller is used to handle the measurable disturbance. It does so by feed-forward cancellation, as can be seen from fig. 16: after a two cycle adaptation the output y_1 is kept on its desired value by a u_1 shaped exactly in accordance with the u_2-oscillations.

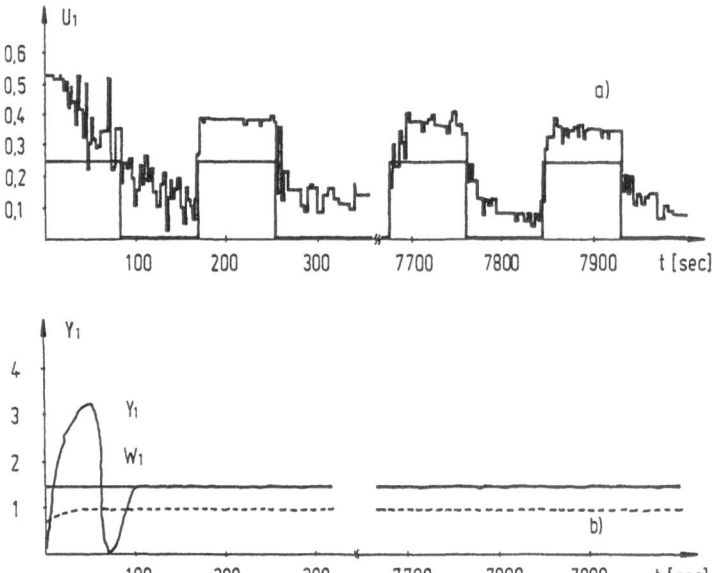

Fig. 16: Learning of measured disturbance feed-forward automatic compensation by LERNAS. a) control u_1 and disturbance u_2; b) process outputs y_1, y_2.

It should be remarked that with the stirred tank reactor the automatic take-over of operator strategies as discussed in II.6 was also demonstrated. On the computer the process was simulated by the equations (24) and brought by a human operator person who had already been trained in this task, from the initial conditions to the considered set point $\underline{w} = (1{,}43\ ;\ 0{,}9225)$. With a sampling time of $T = 0{,}5$ sec and a generalization of $r^* = 8$ the human control actions were stored in parallel in an AMS memory (for details see /11/). Fig. 17 displays the performance of the human operator (curves 1) as well as the performance of the AMS in a repetition of the task (curves 2). One sees immediately that the more critical concentration control is handled much more smoothly by the AMS, certainly due to its inherent generalization, which is equivalent to some smoothing.

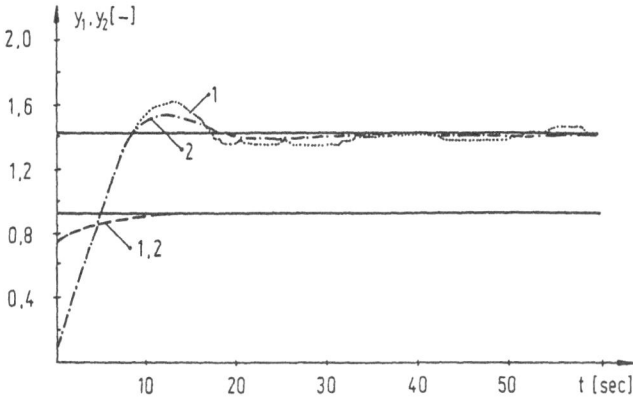

Fig. 17: Transition of stirred tank reactor from initial conditions by a human operator (1) and by an AMS memory which has taken over the operator strategy by being fed in parallel with the situation and his actions (2).

III.1.5.2. Performance criterion variations

It has to be pointed out that one can expect from a learning control loop only those properties for which one has asked by shaping the performance criterion adequately. This can be demonstrated through our stirred tank reactor example. Up to now we have shown mainly the adaptation of the outputs to given set point values. Fig. 18a shows also the course of the control variables in the learned situation. The mass flow of the additional heating especially oscillates very strongly, although the control of y_1, y_2 is satisfactory. Actually, we did not ask for a smooth variation of u_1, u_2 in our performance criterion (26). If additionally we take the control energy into account by an appropriate performance criterion such as:

$$(35) \qquad J(k) = \{[w_1 - y_1(k+1)]^2 + 10\,[w_2 - y_2(k+1)]^2 + 2 \cdot 10^{-5}\,[u_2(k) - u_2(k-1)]^2\}\,T$$

we obtain fig. 18b and we see that now not only the movement in u_2 is reduced to the quantization level but also the movement in u_1. Actually, this cannot be guaranteed, but is a welcome side-effect in our case.

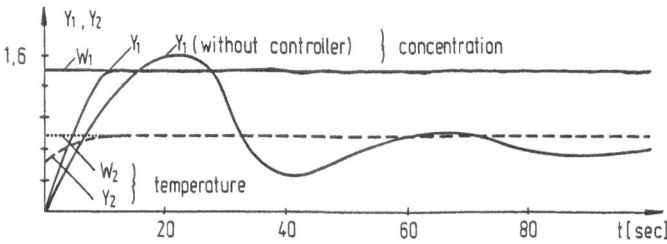

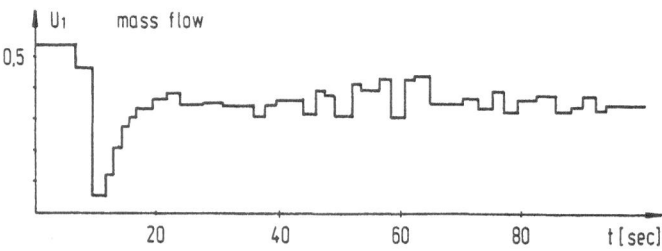

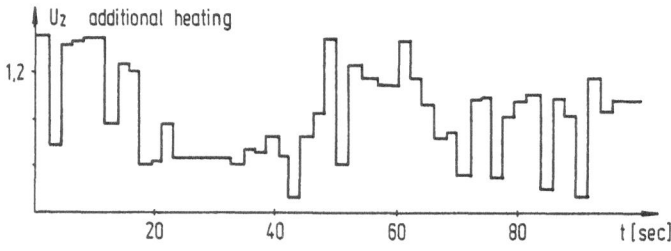

Fig. 18a: Stirred tank reactor with LERNAS in a trained situation; y_1, y_2, u_1 and u_2; performance criterion (26).

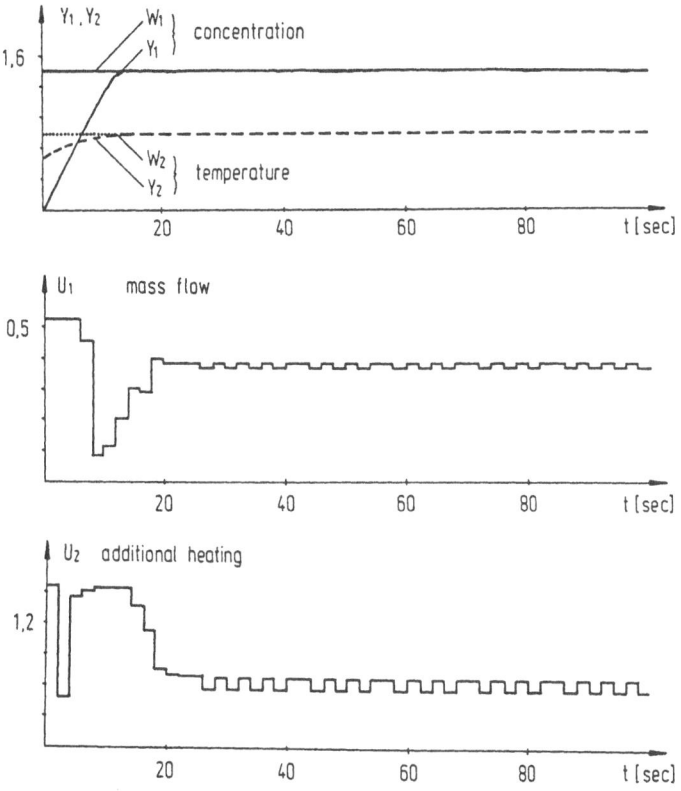

Fig. 18b: Damping down of u_1, u_2 movements to the quantization level by penalizing u_2-differences through the performance criterion (35).

The form of the performance criteria (26) and (34), (35) is still just a consideration of the control differences and/or amount of control by means of quadratic terms. As has been mentioned already in III.1.3. in the context of general remarks regarding optimization this is not necessary. But purposeful changes like the one from (26) to (35) and/or changes in the weighting factors are, in general, of greater influence on the control loop performance than a possible change from squared values to absolute values and/or powers of the order 2ν with $\nu > 1$.

However an interesting improvement is the inclusion of the control deviation tendency: If the tendency is wrong, the control deviation should be weighted more heavily than in the case when it is correct. This can be achieved by distinguishing six cases for $J(k)$ as exemplified in fig. 19 and this has been proven to be very effective in handling processes with backlash and/or friction.

1) $y(k) \geq w$ $\hat{y}(k+1) < w$

$I(k) = A[w - \hat{y}(k+1)]$ $A \gg 1$

2) $y(k) \geq w$ $w \leq \hat{y}(k+1) \leq y(k)$

$I(k) = [\hat{y}(k+1) - w]$

3) $y(k) \geq w$ $\hat{y}(k+1) > y(k)$

$I(k) = A[\hat{y}(k+1) - y(k)] + [y(k) - w]$ $A \gg 1$

4) $y(k) < w$ $\hat{y}(k+1) > w$

$I(k) = A[\hat{y}(k+1) - w]$ $A \gg 1$

5) $y(k) < w$ $y(k) \leq \hat{y}(k+1) \leq w$

$I(k) = [w - \hat{y}(k+1)]$

6) $y(k) < w$ $\hat{y}(k+1) < y(k)$

$I(k) = A[y(k) - \hat{y}(k+1)] + [w - y(k)]$ $A \gg 1$

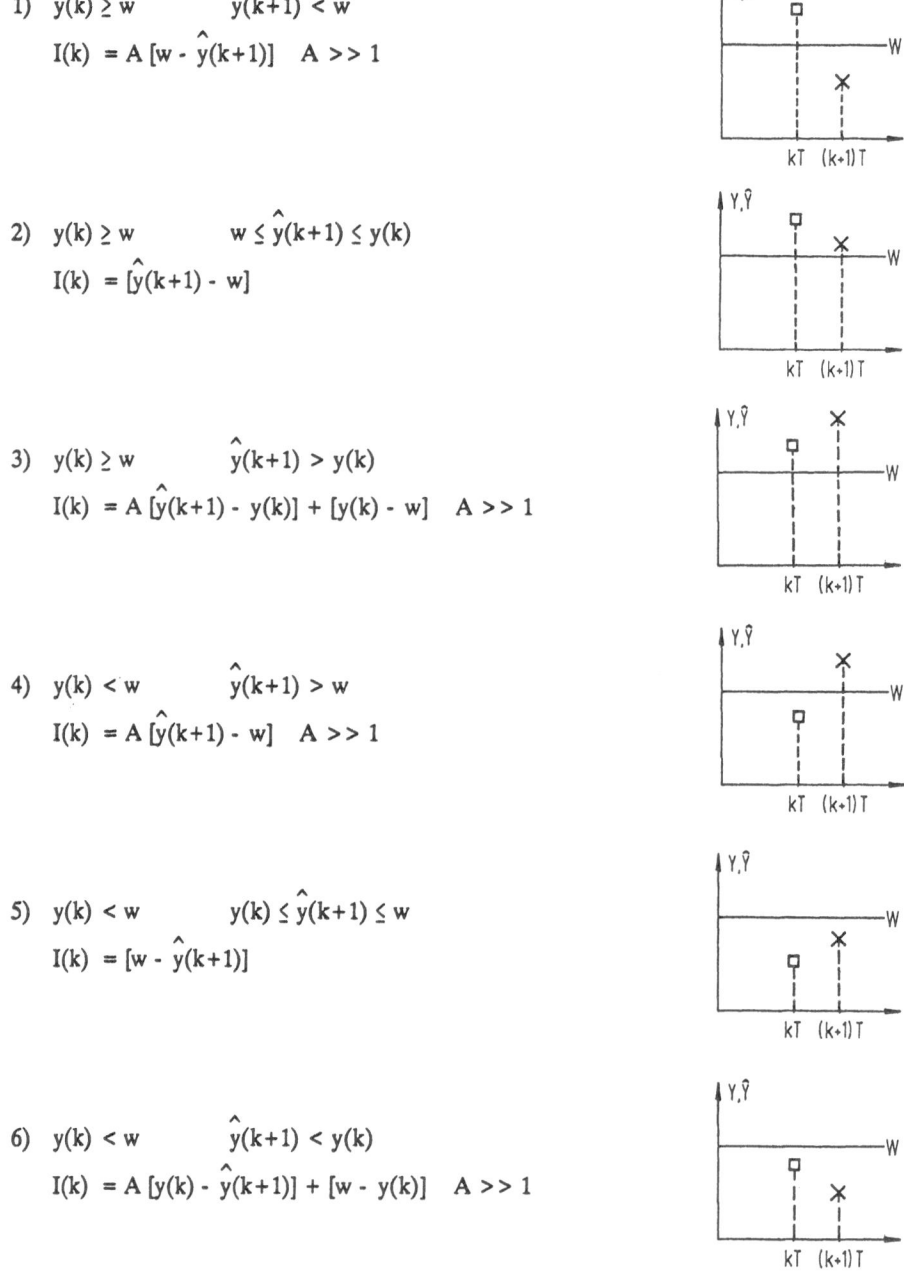

Fig. 19: Performance criterion taking control deviation tendencies into account (w = const, $y(k)$ measured, $\hat{y}(k+1)$ predicted value).

<u>III.1.5.3. Influence of optimization procedure variations</u>

In III.1.3. two optimization procedures have been described explicitly: The Hooke-Jeeves algorithm and a simple evolution strategy. Actually two further algorithms have also been investigated, the Rosenbrock strategy (/9/) and a general evolution strategy (/10/). The Rosenbrock algorithm is a modification of the Hooke-Jeeves algorithm in the sense that, instead of always using the original coordinate directions for the search, now each time one uses a rotated coordinate system, in which one of the axes is pointing in a direction which has been judged to be advantageous in the last step. The general evolution algorithm works not only with one, but with a number of parents and not only with one, but with a number of children, generated now by looking at couples from the parental generation, that means this algorithm mimics much better biological events than the simple evolution strategy. However, both the more complex algorithms did not turn out to be better than the simpler ones - as will be seen from the following discussions - and so they are therefore not described in detail here.

All results to be described have been generated with the help of the stirred tank reactor test process given by equations (24) - (30). Three situations/features have been considered to characterize different cases for the optimization:

(36)   •   Is the process in the transition phase or in the set point keeping phase? As a point of the transition phase, the moment t = 4 sec was chosen, as a point of the set point keeping phase t = 23 sec (cf. fig. 12). The respective marking is:

    T   transition phase point.
    K   set point keeping phase point.

•   Is the optimization algorithm working in some already well trained environment or does it has to cope with an early, badly trained situation with large portions of unexplored process situations? As a well trained case 80 transitions from the initial conditions to set point keeping with an overall run time of 100 sec (50 sampling steps) was chosen. As a badly trained case just the first repetition was considered, that means the situation after one transition from the initial condition to the set point again with a run time of 100 sec and/or 50 samplings. The respective marking is:

    B   badly trained situation.
    W   well trained situation.

•   Is the process model used just once in connection with the optimization of $\underline{u}$ through (26) or is it used twice as the most simple case of a 1-step look ahead procedure which means using a summing up of (26): $\tilde{J}(k) = \frac{1}{2}\left[J(k) + J(k+1)\right]$ as performance criterion and consideration of a four-dimensional optimization problem with $u_1(k)$, $u_2(k)$, $u_1(k+1)$, $u_2(k+1)$ as independent variables? Both possible set ups were taken into account and marked by

    1   one step look ahead set up.
    2   two steps look ahead set up.

With these three distinctions one gets eight different cases, which can be characterized by a combination of the abbreviations.E.g. KB2: means that we are in the set point keeping phase, however, with a so far badly trained process situation and try to optimize with a procedure looking two steps ahead.

To generate informative results the following method was applied:

(37)   • all four different optimization algorithms were run twenty times, started from the same twenty different initial points, randomly chosen in an equally distributed manner.

   • the performance criterion values and the respective $\underline{u}(k)$ and/or $\underline{\tilde{u}}(k) = [\underline{u}(k)',\underline{u}(k+1)]$ were recorded depending on the number of optimization steps n_t and computation time t.

   • mean values with regard to the different starting points were computed depending on the performance criterion and the distance from the global minimum.

The global minimum can be found by complete enumeration, that means a search over all grid-points in the optimization space, since we are working with discretized variables only. However, due to the amount of work, this was restricted to the two dimensional case (one step look ahead set up).

Since the optimization algorithms have certain free parameters, some preliminary investigations were made in advance to compare the algorithms and to tune these parameters to values reasonable for the task considered. However, for the general evolution algorithm this could not be performed completely due to the high number of free parameters. So this algorithm may be partly misjudged.

The algorithms are abbreviated in the following by:

(38)   HJ   Hooke- Jeeves algorithm
           RO   Rosenbrock strategy
           SE   Simple evolution strategy
           GE   General evolution strategy.

The results of the comparison are:

(39)   <u>Convergence:</u>

   • An important aspect of convergence in our problems is whether at first the algorithm reaches at all trained area, since from the randomly chosen twenty starting points some will be outside the trained area in general, which is treated by giving the performance criterion a penalty value. Marking by X the situation in which the algorithm is reaching the trained area, one gets table one. Only the simple evolution strategy is satisfactory in this sense, the others having problems if the situation is not a well trained one.

	TB1	KB1	TW1	KW1	TB2	KB2	TW2	KW2	Σ
HJ			X	X				X	3
RO			X	X				X	3
SE	X	X	X	X		X	X	X	7
GE		X	X	X				X	4

Table 1 - Results on searching after a trained area. X = successful from all 20 randomly chosen starting points. For abbreviations see (36), (38).

- Actually the optimization algorithms converge just to local minima, the randomly working evolution strategies being expected to be more robust against being trapped in such a local minimum. The next step in comparing performance was therefore a comparison of the mean value of performance criteria minima achieved from the different starting points which allow to reach trained area and by that minimization at all. Table 2 gives the relative merits, 1 indicating the lowest mean value, 4 indicating the highest mean value. The Rosenbrock strategy adapting itself best to the local surface structure is trapped mostly in local minima, the simple evolution strategy seems to be best in avoiding being trapped early.

	TB1	KB1	TW1	KW1	TB2	KB2	TW2	KW2	Σ
HJ	3	3	3	3	3	3	2	3	23
RO	4	4	4	4	4	4	4	4	32
SE	1	1	1	1	1	1	1	2	9
GE	2	2	2	2	2	2	3	1	16

Table 2 - Relative performance in reaching low performance criteria values from the different starting points, that means not getting trapped easily by the next local minimum. 4 meaning highest value of mean value of reached performance criteria minima, 1 meaning lowest value.

- A further measure regarding the convergence is, how far the algorithms get to the global minimum. Again the mean values are considered for those optimization runs in which the trained area is reached from the starting point and the optimization is meaningful. Table 3 gives the respective results. However, only the case of a one step prediction is shown, since - as has been remarked earlier - the search for the global minimum by complete enumeration is too costly for the two step ahead prediction set up, due to the fact that this leads to a four-dimensional optimization space. Again the simple evolution strategy seems to be the best algorithm.

	TB1	KB1	TW1	KW1	Σ
HJ	3	4	3	4	14
RO	4	4	4	4	16
SE	1	1	2	1	5
GE	2	2	1	2	7

Table 3 - Mean distance from the global minimum reached in meaningful optimization runs (trained area is reached from the starting point). 4 highest value, 1 lowest value. Two times a 4 was given to HJ and RO in KB1, KW1 since both algorithms reached nearly the same value, however, this value was much worse than the values reached by SE and GE.

(40)   Computation time

- A comparison of the computation time till a minimum is reached is difficult in so far as not all algorithms are reaching the trained area from all starting points - see table 1 -. Therefore only those three cases, for which this is the fact, namely TW1, KW1, KW2 are taken into account here. Fig. 20 shows the respective results. It gives the decrease of the mean value of the performance criterion for all 20 runs from different starting points over the time. There are no numbers shown on the axes. For the mean performance criterion values no real meaning can be attributed to the actual numbers. Also, to show the different minima reached it would have been necessary to use, either a logarithmic scaling or at the end some different scaling at least in fig. 20a, 20b: The value zero is reached only according to the drawing accuracy. With respect to the time one has to take into account that the program used was designed to print out all intermediate steps, so the absolute time necessary has again no meaning. But also the relative times e.g. for KW1 and KW2 minimization are of no significance, since here just one situation is studied, but for the control task, only the amount of time to get the plant sufficiently under control is of major interest.

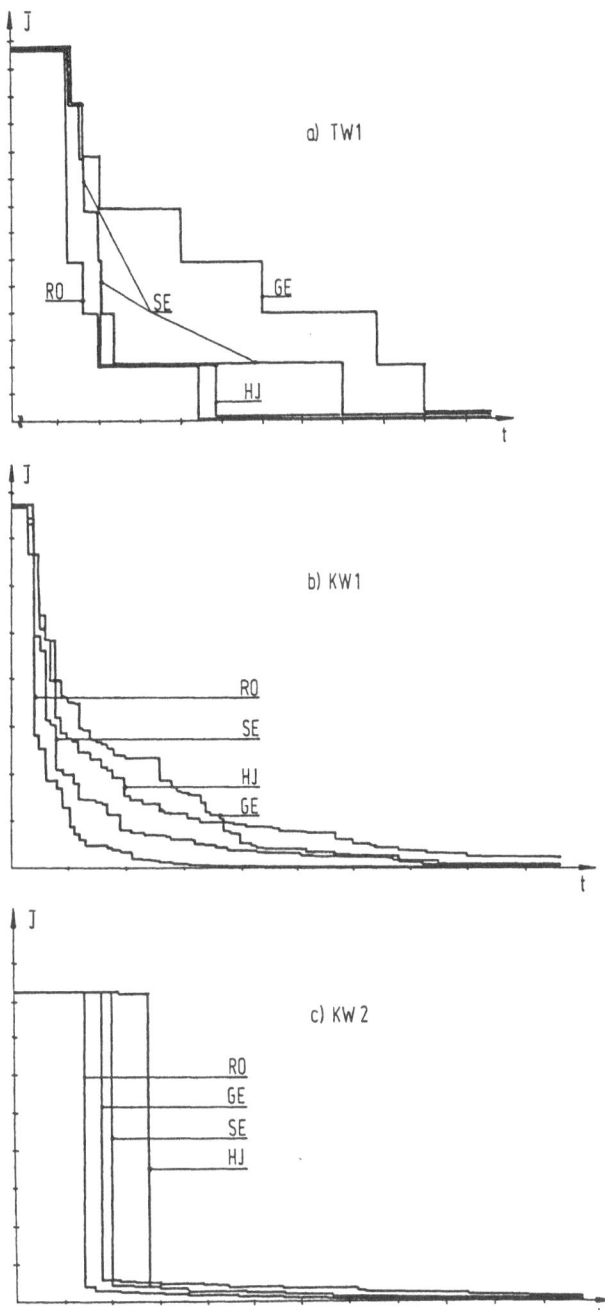

a) TW1

b) KW1

c) KW2

Fig. 20: Decrease of the mean of performance criterion values J reached from different starting points shown over computation time for the four studied optimization algorithms and the cases TW1, KW1, KW2, the only three ones, in which the trained area is reached from all starting points. Absolute values are of no meaning, the relative decrease over time however allows one to judge the computational efficiency of the optimization algorithms. For the meaning of the abbreviations see (36).

Taking the results from the convergence investigations and the relative computation time evaluations together, one comes to the conclusion that the simple evolution strategy seems to be the best optimization approach for the learning control, and that by using it improvements can be achieved compared with the Hooke-Jeeves method chiefly used up to now. But one has to take into account that only one special process has been considered, which is normally not enough for judgements of the optimization methods. Also section III.1.5.4 will show that the selection of the optimization algorithm seems not to be really critical.

III.1.5.4. Optimization improvement limits

In section III.1.5.3. it was stated that the different performances of the considered optimization methods were due to the fact that the methods were more or less finally trapped by local minima. It will be shown here that such local minima exist, and that, furthermore, the topology of the optimization problem is heavily dependent on the training status. This means an optimal control vector generated for a certain situation at an early stage of learning may be completely different from the correct optimal control vector for the given situation with an exact plant model.

For the respective computations the stirred tank reactor with the performance criterion (26) was again used. The problem of set point keeping has been considered. At first the global minimum was determined by complete enumeration taking into account the exact equations (24). This global minimum was then selected as the centre point of charts with the coordinates $u_1(k)$, $u_2(k)$ and lines of equal performance criterion value indicating the topology of the surface, in which the minimum search is performed. Fig. 21 shows the respective charts for three cases: a) gives the surface stemming from the exact equations, b) shows the surface after 100 repetitions of going from the initial conditions to set point keeping, c) represents finally the situation at the first repetition. One sees that after 100 runs the topology is in general satisfactory, but that the global minimum is still not exactly achieved and that furthermore a number of local minima exist. The situation is much worse after the first run: here not even the general topology is represented.

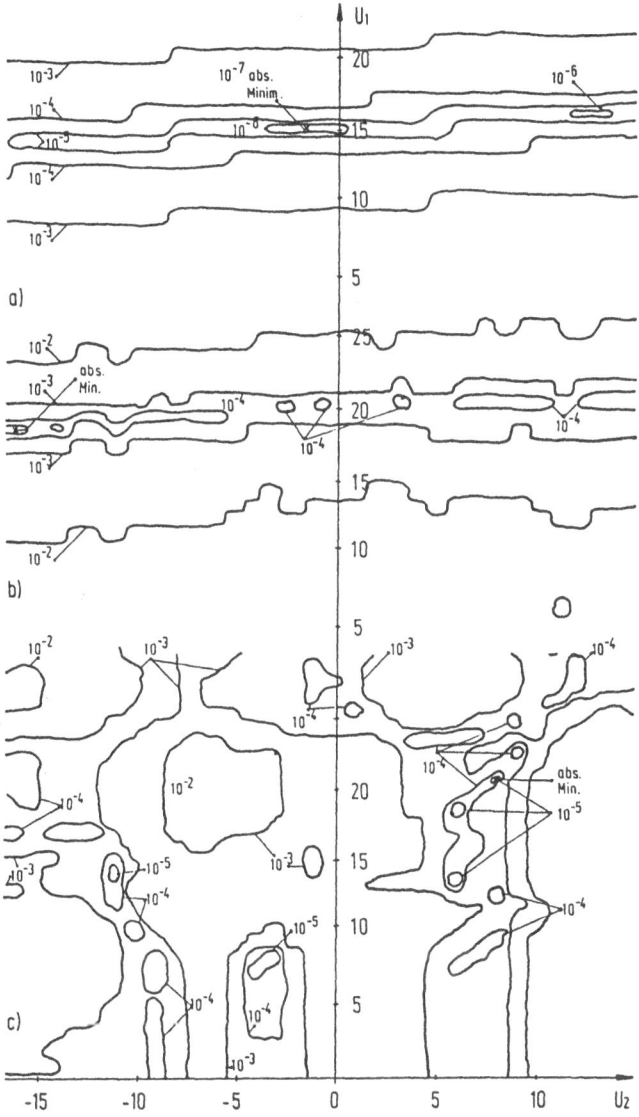

Fig. 21: Lines of equal performance criterion value, describing the surface over $u_1(k)$, $u_2(k)$ in which the minimum has to be found. Situation: Set point keeping a) ideal case, b) learned in 100 runs, c) learned in 1 run.

Fig. 22 shows furthermore that the topology and/or topology change is dependent of the AMS parameters. It displays the same chart as fig. 21 b) but for an AMS with $r^* = 16$, since for fig. 21 $r^* = 32$ was used.

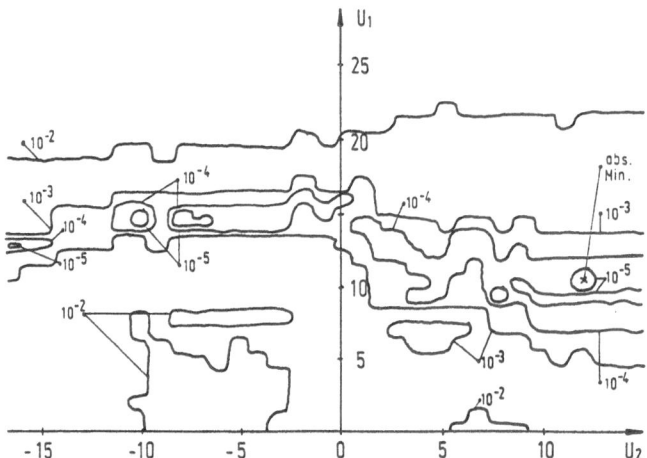

Fig. 22: Same situation as in fig. 21b), however generated with an AMS of $r^* = 16$ instead of $r^* = 32$ as in fig. 21b).

The result of these investigations is twofold:

First: Since in all experiments and applications of LERNAS the learning control loop was able to control the plant in a satisfactory way, the overall control loop structure must be fairly robust against wrong setups in the topology of the optimization problem.

Second: Since the topology during learning is so different from the real topology, it is not very important which optimization procedure is used. It does not make sense to put a lot of work into research in this area.

It should be further added that in real time control - which will be discussed later on - the optimization process does not end by finding a minimal value with a certain accuracy but is limited by computation time. This means only some steps in the direction of a minimal value will be performed. However, this seemed to be acceptable since the main optimization gains are always achieved in the first steps - see also fig. 20 -. Our findings here are supporting such a line of action since the topology in each learning cycle and $\underline{u}^{opt.}$ are changing anyhow for a given situation.

Consequently it is not of major importance, what sort of approximation to $\underline{u}^{opt.}$ is stored away for further use at a given moment of the learning control loop development.

III.1.6. Robustness with respect to selectable parameters

III.1.6.1. Method of assessment

LERNAS has a number of parameters which have to be fixed before it can be used. Looking at e.g. (30) one sees that these are:

(41)　　　　Sampling time T
　　　　　　Training indication value η
　　　　　　Quantization ϵ
　　　　　　Generalization r^*
　　　　　　Optimization default value h_0 (Hooke-Jeeves)
　　　　　　Aktive learning step length b_0.

Since the aim of LERNAS is to handle relatively unknown, complex processes it is important that it is not necessary to tune these parameters very accurately to achieve good performance.

Certainly something has to be known about the process. There should be some rough idea about sampling time, quantization and range of variable variations (to make enough memory capacity available) as well as knowledge about the questions of how far the measurements (measured values $\underline{y}$, $\underline{v}$) suffice to give information about all major effects regarding process behaviour (observability, major disturbances) and how far the actuation possibilities (inputs $\underline{u}$) suffice, to influence the major process properties (controllability). However, this is normally the case for engineering problems and is an unavoidable requirement to control processes.

Thus the remaining question is how far fine tuning is of importance, having made rough estimates. Since memory space is now relatively cheap a certain overdesign in this respect seems to be acceptable, so that the parameters listed in (41) are left for investigation.

Again the stirred tank reactor process as described in (24) - (29) was used. The transition from the initial values given in (25) to the set point given in (27) was considered in k_1 sampling steps just n_1 times. Now a first task was to find adequate criteria with which to compare the performance of LERNAS depending on parameter variations.

A first criterion is certainly the minimum performance criterion value being reached by learning. To avoid misunderstandings it has to be pointed out that in III.1.5.3. we considered the performance criterion values during the optimization in a given situation, whereas we are now considering the performance criterion values reached during the full transition from initial values to set point keeping, being optimized for each individual situation. We characterize this performance criterion value by the number of the previously considered transition steps and we use the sum of the individual optimized values:

$$(42) \qquad J_n = \sum_{k=0}^{k_1-1} J(k)$$

The minimum performance criterion value used as the first criterion is now:

$$(43) \qquad J_{min} = \min_n J_n \; ; \qquad\qquad n = 1,2 \dots n_1$$

Actually we know from learning curves like fig. 13 that for big enough values of n_1 one is in a monotonic decay interval and that one reaches a good approximation to J_{min} as $n_1 => \infty$.

A second point of interest is how long it takes to reach J_{min}, that means, how steep is the learning curve. A criterion to assess this, proposed by J. Militzer in /11/, is:

$$(44) \qquad N_L = \frac{\sum_{n=1}^{n_1} (J_n - J_{min})}{J_o - J_{min.}}$$

One sees immediately that if $J_1 = J_{min}$ already and all further J_n stay equal to J_{min}, N_L delivers the value zero and that with slower decay of J_n N_L becomes larger and larger.

As a possible criterion to assess the stability of learning, the monotony of decay may be used. For this Militzer proposed:

$$(45) \qquad M = \frac{\sum_{n=1}^{n_1} | J_n - J_{n-1}|}{J_o - J_{min.}}$$

A monotonic decaying learning curve has the minimal value of $M_{min} = 1$, while for a non-monotonic decay one gets $M > 1$. (45) is a relatively sharp criterion, since just one major deviation from the monotonic decay resulting from active learning may greatly change the value of M. Therefore variations in M should not be looked at too critically.

For further information on the motivation for the criteria (43) - (45) see /11/. There one can find also some discussion of why and in which sense the selected criteria are different from the criteria proposed in /12/ by G.N. Saridis to assess learning.

The AMS memories work using hash coding and the hash-collisions thus produced are only satisfactory if the memories are not filled up more than 80 % - 90 %. Since the development of the control strategy certainly depends on the accuracy of the predictive model learning behaviour, the control quality will be diminished if the model storing AMS is used to too great an extent. Therefore

(46) A_M the filling grade of the AMS storing the predictive process model, taken at the end of the learning runs: $n = n_1$

is displayed in addition to the criteria (43) - (45).

III.1.6.2. Results

For all the results

- a time interval $0 \leq t \leq 100$ giving the respective value k_1 for (41) and a number of $n_1 = 50$ learning runs

has been used.

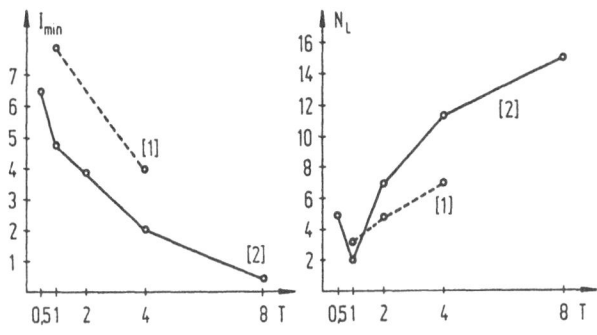

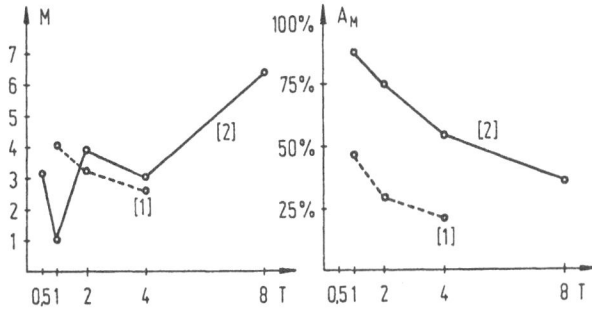

Fig. 23:   Sampling time dependence for

(1) $\eta = 80\ \%,\ \epsilon = 1\ \%,\ r^* = 8,\ h_o = 8,\ b_o = 4$

(2) $\eta = 80\ \%,\ \epsilon = 0,5\ \%,\ r^* = 16,\ h_o = 16,\ b_o = 8$

Fig. 23 shows the influence of a change of the sampling time T on the criteria (43) - (45) and A_M for two sets of the other parameters. However, the results should not be misinterpreted: J_n is summing the squares of the control deviations at the sampling points. For a short sampling time a higher number of sampling points fall into the transition phase than otherwise, and in the transition phase we have high control deviations leading to the major part of the performance criterion value. Therefore one gets higher values of J_{min} for shorter sampling intervals, although - as has been found out by additional computations - the integral of squared control deviations is nearly independent of the chosen sampling time. Only for the sampling time of 0,5 sec the integral of squared control deviations is getting considerably larger, however this is due to the too high process model memory filling, as can be deduced from the display of A_M. Since in the following comparisons the problem explained above does not exist but computing of the integral of squared control deviations is relatively costly, J_{min} was not replaced by the integral of squared control deviations as an alternative criterion.

The interrelation between N_L, A_M and T is due to the fact that with more frequent sampling one obtains more information during each transition and consequently a good process model training with respective optimization success appears more quickly. The monotony index does not show very much variation, except in one case.

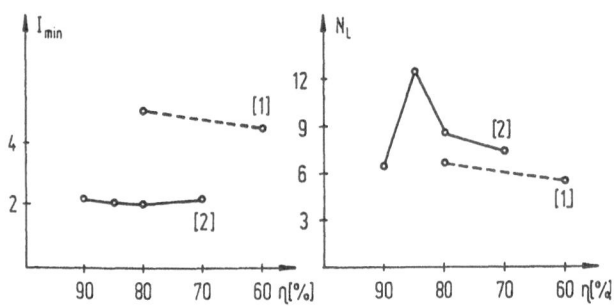

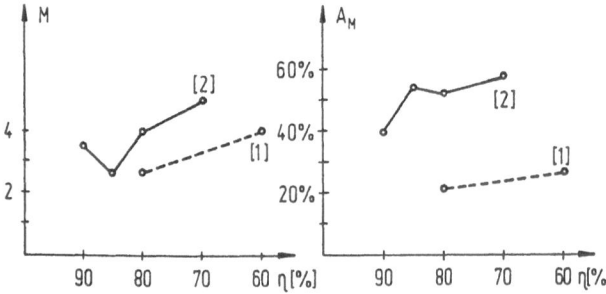

Fig. 24:   Trainingsindicator dependence for

(1) $T = 4$ sec, $\epsilon = 1\%$, $r^* = 8$, $h_0 = 8$, $b_0 = 4$

(2) $T = 4$ sec, $\epsilon = 0,5\%$, $r^* = 16$, $h_0 = 16$, $b_0 = 8$.

From fig. 24, which comprises the results of training indicator variations, one can deduce that this parameter is in general not critical. Only for one value $\eta = 85\ \%$ one gets somewhat differing values.

Fig. 25 displays the result of quantization changes. J_{min} shows - with the exception of $\epsilon = 0{,}33\ \%$ where again A_M is too high - a remarkable independence from ϵ. With growing ϵ one gets a greater amount of generalization and consequently a less necessary memory capacity as well as an acceleration of learning. The improvement in the early predictive process knowledge (less untrained area in the case of not so fine quantization/growing ϵ) leads at the same time to faster learning and higher monotony in the learning curve. For the curves for parameter set (1) one has to take into account that the number of association cells and consequently the generalization is only half as big as for parameter set (2).

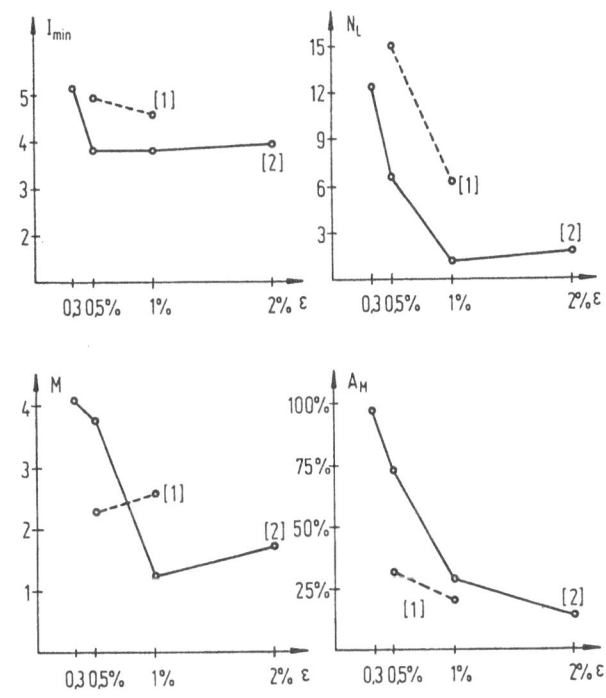

Fig. 25:   Quantization dependence for

(1) $T = 4$ sec, $\eta = 80\ \%$, $r^* = 8$, $h_0 = 8$, $b_0 = 4$

(2) $T = 4$ sec, $\eta = 80\ \%$, $r^* = 16$, $h_0 = 16$, $b_0 = 8$.

Although the product $\epsilon \cdot r^*$ gives the overall generalization it is a different matter whether one changes the quantization or the number of cells among which a certain value is distributed: r^* gives the amount of interpolation/smoothing as one can conclude easily from $r^* = 1$, where no interpolation happens at all. Fig. 26 puts forward the dependence on r^*. Interesting to note is the unsatisfactory behaviour for $r^* = 8$, which is however, in accordance with the results in fig. II.15. A detailed inspection of the respective test runs show that with $r^* = 8$ the number of transitions used here ($n_1 = 50$) is not sufficient. Due to the relatively low local generalization one obtains by active learning even during the last 5 transitions still in over 80 % of the sampling, steps into untrained area in the neighbourhood of the estimated optimal input $\hat{u}^*$ (cf. fig. 3). With all other r^*-values one has only a small influence on the control performance J_{min}. N_L and A_M decay with growing generalization as would be expected. Also an improved stability of learning takes place, since gaps in the trained area are closed faster by growing generalization.

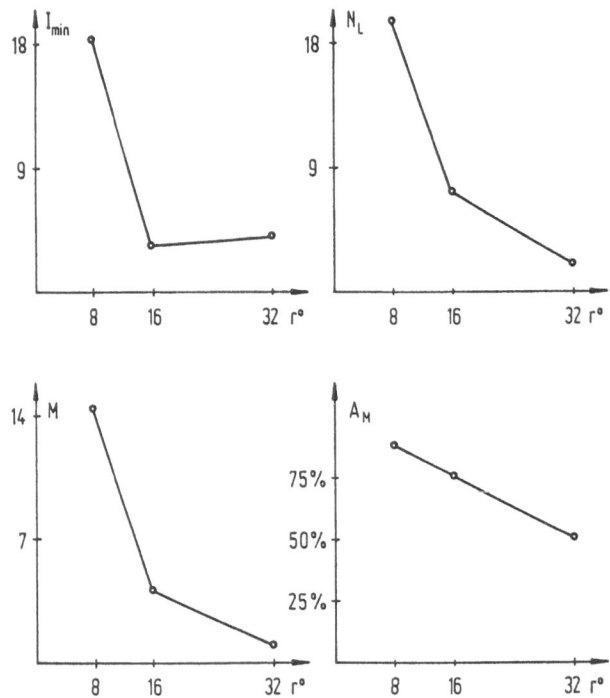

Fig. 26: Dependence from interpolation factor r^* for $T = 2$ sec, $\eta = 80$ %, $\epsilon = 0,5$ %, $h_0 = 16$, $b_0 = 8$.

The variation of the default value (initial value) of the step size in the Hooke-Jeeves algorithm is of minor influence as can be seen in fig. 27. As may be anticipated with growing h_0 one obtains some reduction in learning duration N_L, for which one has, however, to pay by roughly 10 % additional computation effort at each sampling step.

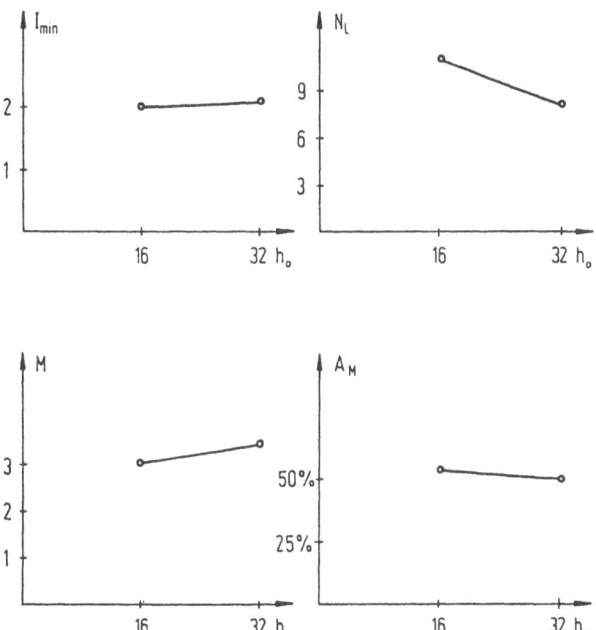

Fig. 27: Dependence of optimization step size (Hooke-Jeeves algorithm) for T = 2 sec, η = 80 %, ϵ = 0,5 %, r^* = 16, b_0 = 8

With reduced active learning (b_0 = 0 or only a small b_0) in the framework of the given amount of transitions the system does not obtain enough information about the process behaviour to allow an effective optimization of the control strategy to take place. This expresses itself also in a very low memory use (A_M). With growing b_0 more and more information is gained and the optimization becomes effective and J_{min} decreases (see fig. 28). The stability of learning decreases also, as would be expected, since b_0 is designed to lead into unexplored situations. It is remarkable that the course of learning is not monotonic even in the case of b_0 = 0, M = 2. This is due to the extrapolation of the known area by local generalization even for border points of the trained area, especially for the first steps of learning, when no trained area exists at all.

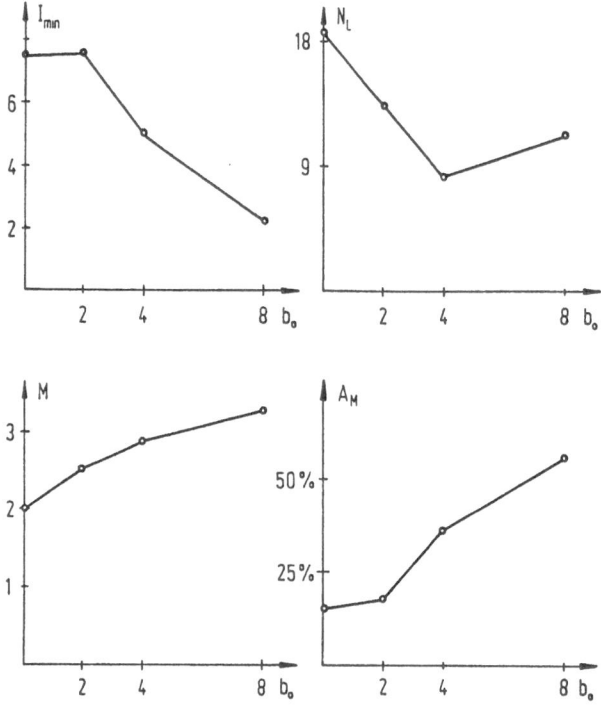

Fig. 28: Dependence of the value for exploration of unknown area (active learning) for T = 4 sec, η = 80 %, ϵ = 0,5 %, r^* = 16, h_0 = 16

In conclusion one can state that the achievable control performance has been found to be relatively insensitive to the parameter fine tuning, at least for the test set up considered. This can be interpreted as the ability of the learning control loop to adapt itself not only to the process and the performance criterion but also to the selected parameter values. The biggest influences on the achievable control performance are the choices of the generalisation/interpolation factor r^* and the active learning step width b_0. Changes in the parameters to be chosen express themselves mainly by changes in the duration and termination of learning. An adaptation to an disadvantageous parameter selection leads to a higher learning effort. The necessary memory space is also essentially dependent from the parameter selection.

III.1.7. Experiments with variable generalization

In II.1.5. we discussed the idea of using not just a fixed generalization but some variable generalization, either by using AMS memories with a different amount of generalization in parallel and selecting for an given situation just that one which has the smallest generalization but yet is still characterized as trained for the situation by the training indicator, or by using the MIAS memory with its data driven, self-adapting generalizing ability. We have mentioned also already that variable generalization ist most important in cases where different degrees of training and different requirements regarding accuracy are appropriate for different situations. In control problems the transition from initial values to a set point gives less data but is also less important with respect to final performance than the set point keeping phase.

It was therefore only natural to look at the results of variable generalization used in LERNAS. Since - as we shall see - LERNAS worked well with AMS and MIAS (showing improved results vis-à-vis the case of fixed generalization) these experiments demonstrated also that the LERNAS concept is not dependent on the choice of the memory systems used for storing the predictive process model and the optimized controller. This means it is appropriate to look on microstructures and macrostructures separately, as has been done in this book.

As a test example the stirred tank reactor was again used, being a multivariable, strongly nonlinear process and thus a really demanding control task. The equations used are (24), (25) and (27) - (29) from section III.1.5.1. The considered set point - equation (27) - is interesting since the process tends to only weakly damped oscillations there, which makes it difficult for an untrained human operator to handle the transition from the initial values to this set point and to maintain it.

The experiments (reported for the first time in /13/) were made with a modification of the LERNAS program, which has been developed for real time, on-line application and which will be explained in the context of the discussion of LERNAS' handling of pilot processes not simulated but existing in hardware. However, the modification comprises only slight changes in the order of the data handling, so that it has no major influence on the LERNAS performance. The use of this version was only mentioned for the sake of completeness.

Since it was shown in the preceding sections that mainly the sampling time T and the generalization properties are of importance, we shall give just the respective figures.

The sampling period was 2 sec in all cases.

Three different memory configurations were used:

(47a)   A fixed AMS (AMS-f) with $r^* = 16$, $\epsilon = 1$ % as well for the predictive process model as for the controller.

(47b)   Three AMS modules organized in accordance with fig. II.20 for variable generalization (AMS-v) with

$r^* = 16$, $\epsilon = 0,5$ %   module 1

$r^* = 16$, $\epsilon = 1$ %   module 2

$r^* = 16$, $\epsilon = 2$ %   module 3

again as well for the predictive process model as for the controller.

(47c)   MIAS with $q = 9$ for the predictive process model and $q = 6$ for the controller.

With regard to (47b) one could argue whether not a fixed ϵ, e.g. $\epsilon = 1$ % of the range of values, and a variation of r^*, e.g. $r^* = 8, 16, 32$, would be an interesting alternative. However, r^* is to be considered not only as an interpolation and/or smoothing factor but also as giving the amount of gradation in the basic local functions - see e.g. fig. II.8. The chosen solution considers this fact by choosing an acceptable gradation and varying the steepness of the flanks of the curves thus approximated by changing the size of the basis by means of ϵ.

The first experiment considered the case of the transfer of a human operator strategy to a locally generalizing memory as controller, that means the situation of fig. II.29 and/or of fig. I.5 without the learning loop being active. Two somewhat different transitions performed by a trained human operator and shown in fig. 29 (indicating by the curves for $y_1 = x_1$ that it is not an easy task, as mentioned above) were transferred into the controller memory. The problem considered here was:

(48a)   To repeat the task for the same initial conditions

(48b)   To repeat the task for $\underline{x}(0) = (0,362; 0,733)'$ that means with a change of the initial concentration by roughly 12 % of the respective range.

(48c)   To repeat the task for $\underline{x}(0) = (0,000; 0,807)'$ that means with a change of the initial temperature by roughly 12 % of the respective range.

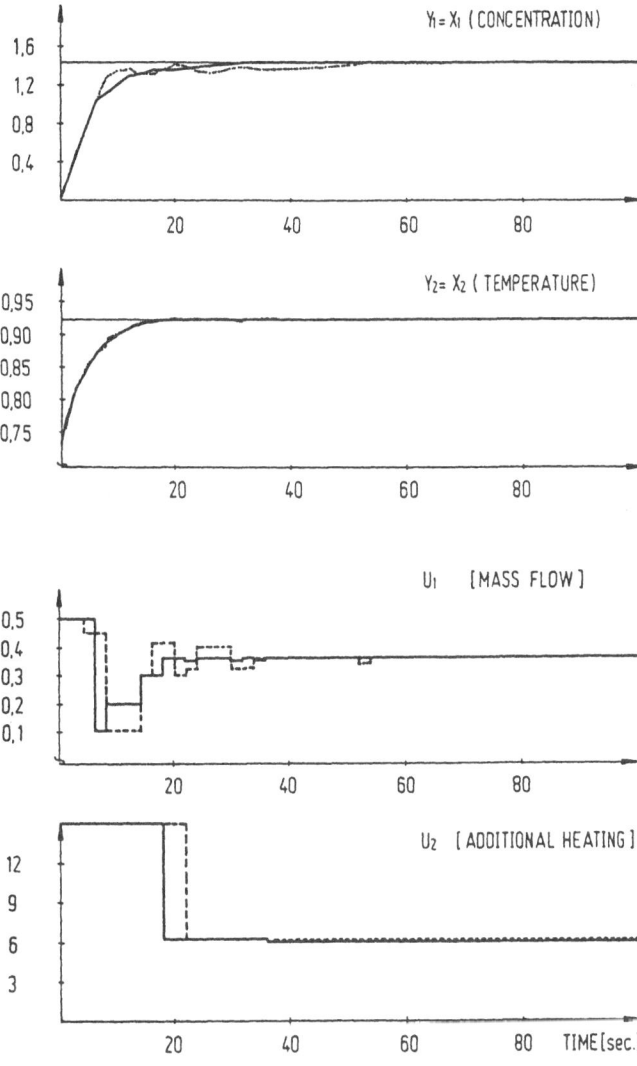

Fig. 29: Trajectories from human operator as basis for locally generalizing memory different set ups (47) performance comparison.

The results are displayed in fig. 30, 31 and 32. For all three cases the repetition (48a) can be performed in an acceptable way. However, the fixed AMS is not able to solve task (48b), (48c): The control variables remain zero and the process keeps its initial status. The variable AMS solves the problem, but for (48b), (48c) does not reach the demanded value w_1. MIAS however adapts its strategy and consequently the transition trajectories in a fully adequate way.

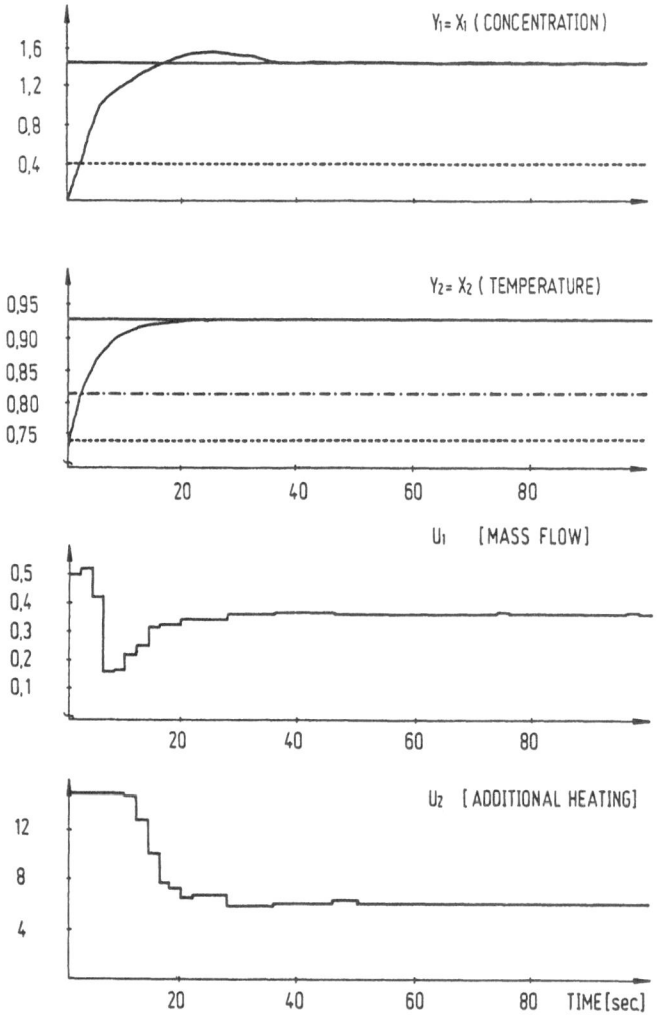

Fig. 30: Repetition and trials to solve (48b), (48c) after training according to fig. 29 by use of AMS-f ((47a)). - repetition (48a), --- task (48b), - · - ·· task (48c).

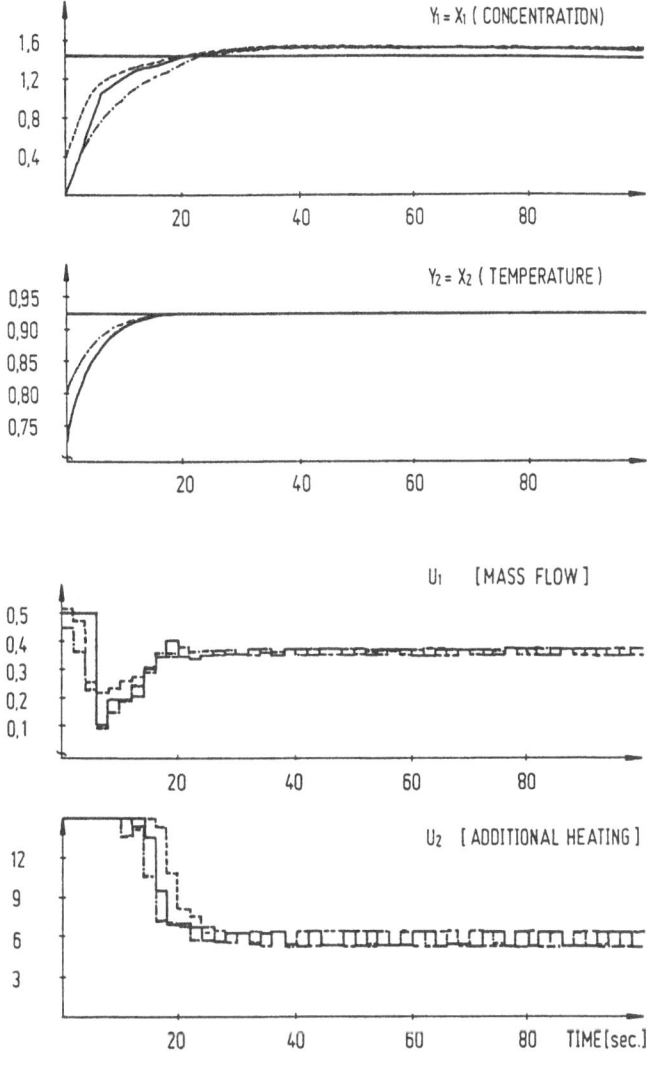

Fig. 31: Repetition and trials to solve (48b), (48c) after training according to fig. 29 by use of AMS-v ((47b)) - repetion (48a), --- task (48b), - ·· ·· task (48c)

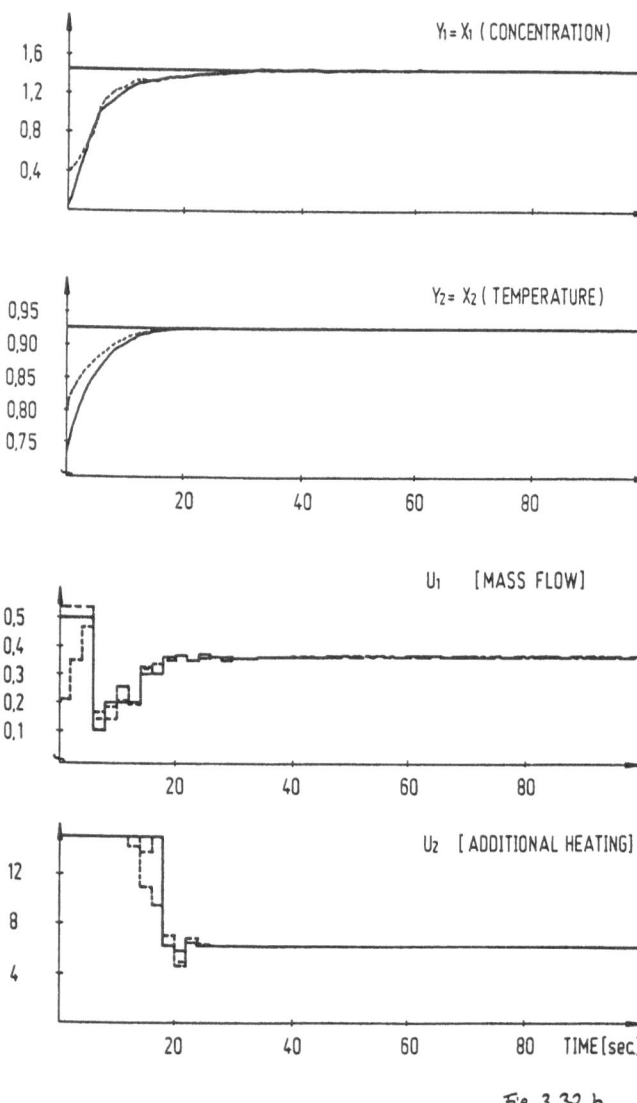

Fig. 32: Repetition and trials to solve (48b), (48c) after training according to fig. 29 by use of MIAS ((47c)) - repetition (48a), --- task (48b) - · - ·· task (48c)

For testing LERNAS itself a performance criterion somewhat different from (26) was used, namely the three step ahead looking criterion:

$$(49) \qquad J(k) = \sum_{j=1}^{3} \left[\frac{|w_1 - \hat{y}_1(k+j)|}{y_{1max} - y_{1min}} + \frac{|w_2 - \hat{y}_2(k+j)|}{y_{2max} - y_{2min}} \right]$$

which also demonstrates the great freedom one has in choosing performance criteria for a given task.

The problem considered was to optimize the transition after preliminary training with the four trajectories shown in fig. 33 performed by a human operator.

The results of the optimization are shown in fig. (34) - (36) by displaying examples from the respective runs. For alle three memory types one has in the beginning deterioriation of control performance. This is due to the low level of pretraining by just four transitions especially in connection with the three step look ahead performance criterion. The information existing in the memories is not sufficient for genuine optimization. Application of control actions not safeguarded by process model knowledge forces the process into unforeseen situations especially in view of its inclination to oscillate. However, this "experimentation" obtains for the system the missing information and is thus able to improve its behaviour in the subsequent runs.

For the fixed AMS one can deduce from fig. 34 that the optimization is not yet brought to an end with the 12th run: the states show still some overshoot and the inputs have not reached stationary values. For variable AMS - fig. 35 - we have a clearly improved behaviour, although the inputs move still in an uncertain way. Finally for MIAS we get the best results - see fig. 36 -. For $y_1 = x_1$, $y_2 = x_2$ we have a fully satisfactory transition to the set point values w_1, w_2 coupled with a smoothing of the inputs mostly to the unavoidable level of one quantization step (which can be suppressed e.g. by some additional nonlinearity). The smoothing of the inputs is of interest insofar, that this was possible with a one step look ahead criterion only by including the amount of change of u_2 into the criterion (compare III.1.5.2.). Furthermore one finds from fig. 36, that the data driven variable generalization from MIAS accelerates the performance optimization vis-à-vis the graded variable generalization of AMS-v, since only four optimization runs are necessary here. Also the time to reach the set point values is in fig. 37 less than that in fig. 36.

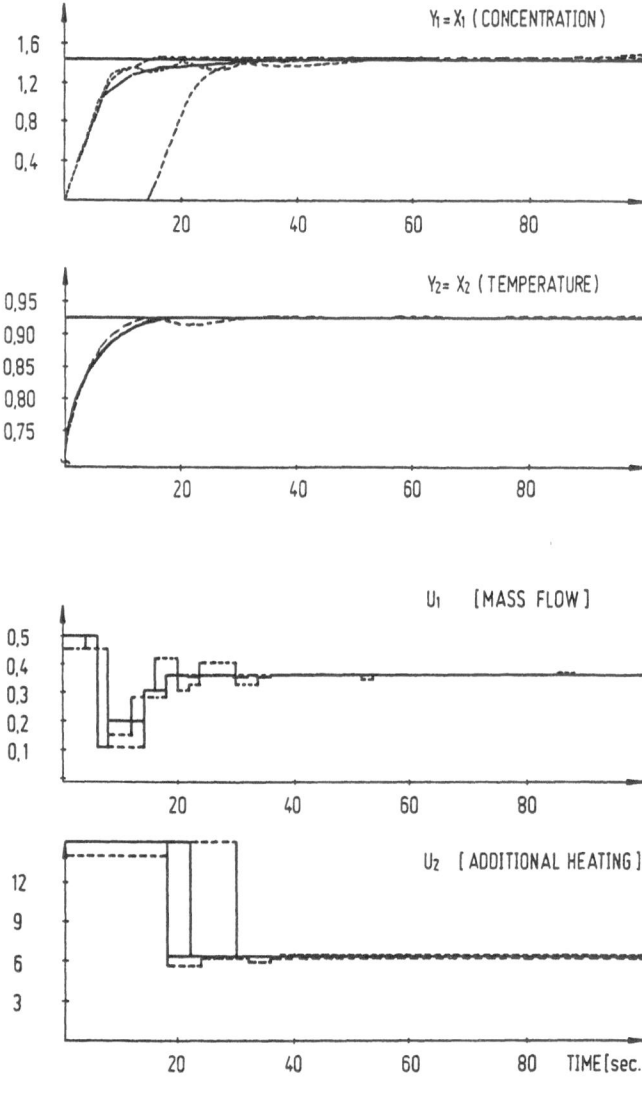

Fig. 33: Trajectories from a human operator used as basis for transition optimization by (47a) - (47c).

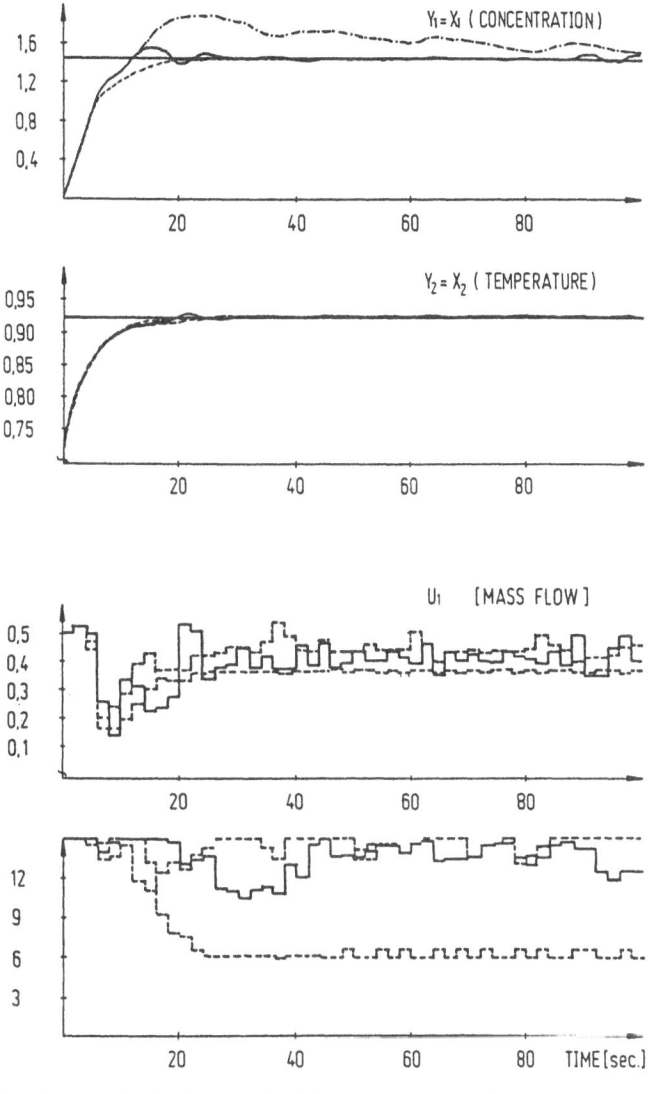

Fig. 34: Optimization on the basis of trained knowledge from fig. 34. AMS-f ((47a)), performance criterion (49): · · · without optimization; - ·· 3rd optimization run; - - - 12th optimization run

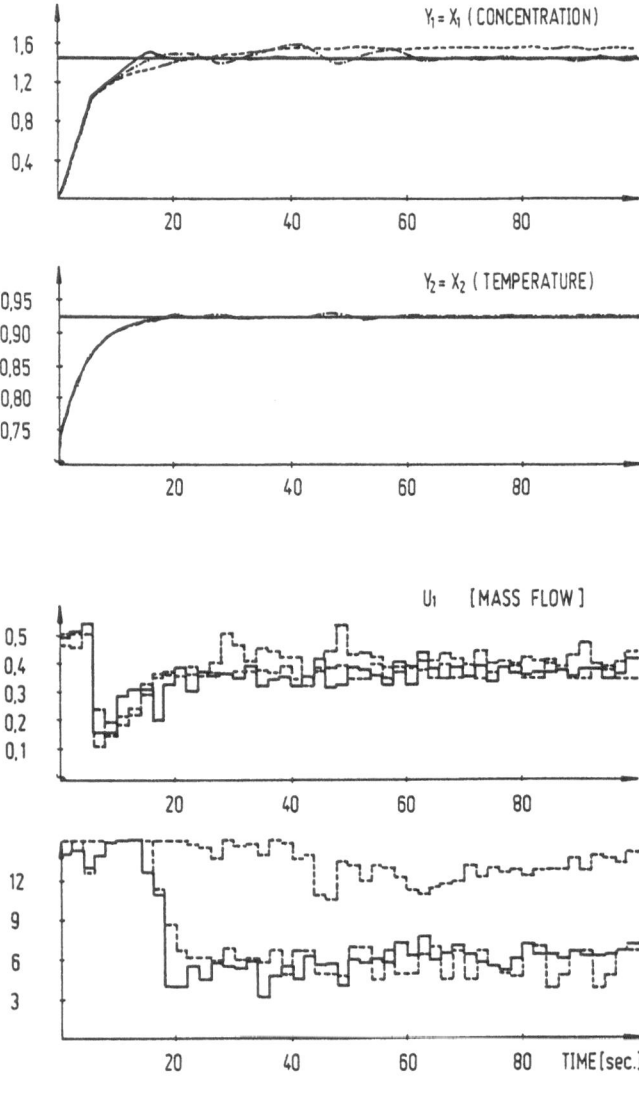

Fig. 35: Optimization on the basis of trained knowledge from fig. 34. AMS-v ((47b)), performance criterion (49): ··· without optimization, -··- 3rd optimization run, --- 12th optimization run.

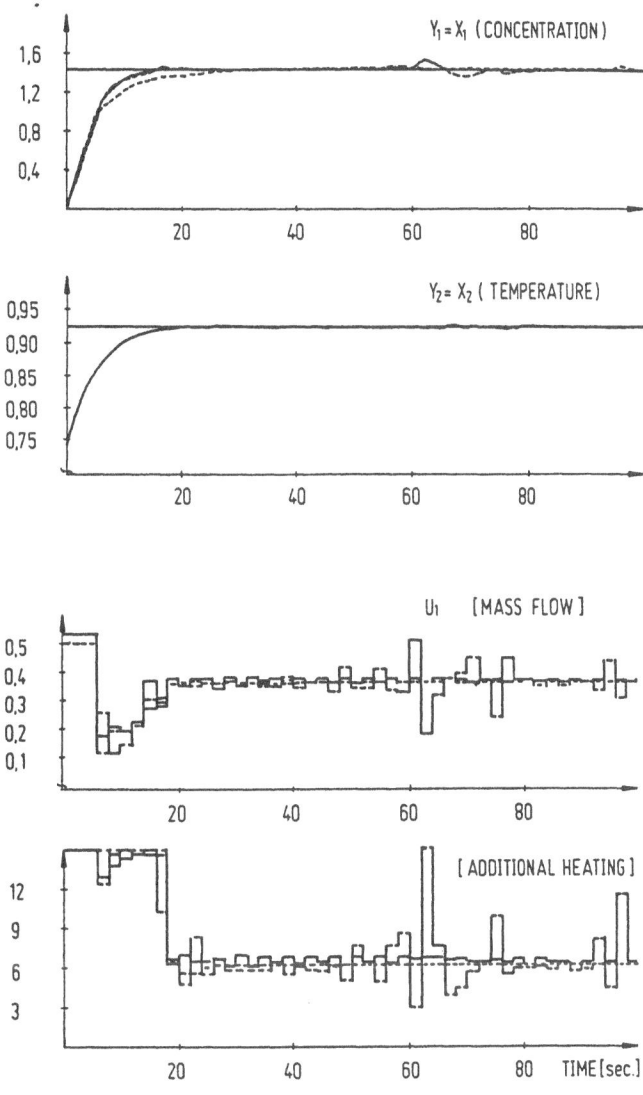

Fig. 36: Optimization on the basis of trained knowledge from fig. 34. MIAS ((47c)), performance criterion (49): ··· without optimization, - ·- 2nd optimization run, --- 4th optimization run.

As a conclusion one can state, that variable generalization seems to be a helpful tool in learning control loops. The data driven variable generalization of MIAS performs best. However, if not just the taking over of some experts process control strategy with a limited amount of training points is considered, but rather a selfoptimizing system like LERNAS, one has to take into account that a steadily growing number of training points has to be handled and that the response time of MIAS is dependent of the number of training points to be searched through. Thus for LERNAS AMS-v may be best and for the automatic taking over of operator strategies MIAS.

III.1.8. Supplementary remarks

III.1.8.1. Closed loop stability

For most plants it is of the utmost importance that no excessive process inputs and/or outputs and/or high frequency oscillations are generated by operating them in a closed loop. Therefore stability considerations are an important aspect in system theory.

However, one has to take into account that system theory is working with idealizations, that is plant models, which are more or less simplified descriptions of reality. Therefore mathematical stability proofs do not guarantee in general that the selected controller leads to a stable control loop when applied to the real plant. A practical way out is continuous monitoring of the plant behaviour and the use of safety measures when certain critical limiting values are exceeded.

Basically, this seems to be the appropriate strategy for LERNAS, too, since learning control loops make most sense, if applied to processes, about which no detailed knowledge is available.

However, some principal results can be established from global considerations and simulations.

First: LERNAS is BIBO stable for BIBO stable processes. This is due to the following fact: For bounded-input /bounded-output (=BIBO) stability one obtains by certain simplifications from /14/ as a definition: A time-discrete multi-input/multi-output system is BIBO stable, if with a finite real number M for all input signals $\underline{u}(k) \in \mathbb{R}^{n_u}$ fulfilling the condition $|u_i(k)| \leq M$ (i = 1,2...n_u; k ≥ 1) some finite real number P exists, so that for the output signals $\underline{y}(k) \in \mathbb{R}^{n_y}$ one has always $|y_j(k)| \leq P$ (j = 1,2 ... n_y; k ≥ 1). Now, $\underline{u}$ is from a finite region in any case in LERNAS, which leads for a BIBO stable process to a limited process and closed loop output $\underline{y}$ and consequently to BIBO stability of the closed loop itself also. Criteria for a test regarding BIBO stability may be found also in /14/.

Second: With respect to handling of non-BIBO stable processes by LERNAS we have studied some very simple plants with integrating properties:

(50) $$y = \frac{k}{s \cdot (1+T_s)} \cdot u = \frac{0,25}{s(1+1,125s)} \cdot u$$

by simulation.

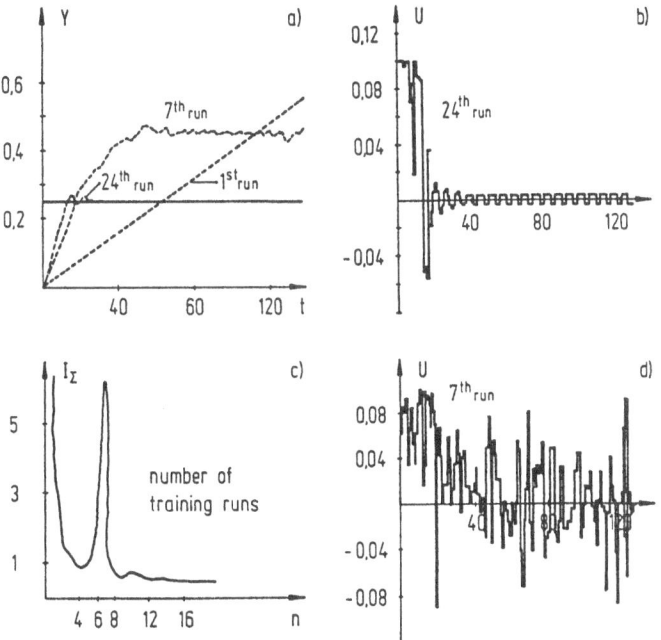

Fig. 37: Control of a simple unstable process by LERNAS: a) w(k) = w = const = required output and outputs of the first run (u=o), the 7th and 24th run. b) Process input behaviour for the 24th run. c) Performance criterion value in dependence of the number of runs, deviation at the 7th run due to active learning. d) Process input behaviour for the 7th run.

Fig. 37 shows a fairly typical result from a series of test runs with different values of the free parameters in LERNAS, such as the sampling time T, the amount of local generalization r* and so on. One obtains the following impression: LERNAS is also able to handle unstable processes - see fig. 38a -. However, active learning may cause large deviations from the desired behaviour - fig. 38c, d - and the performance seems to be more sensitive with regard to the selection of the freely adjustable parameters in LERNAS.

III.1.8.2. Backlash, friction, dead- time and non- minimum phase plants

High nonlinearity due to exponential relationships as displayed in the pH- control and stirred tank reactor processes is not the only difficulty encountered in control system design. Problems stem also from discontinuous behaviour due e.g. to friction or backlash, from long time delays and from plant responding in an opposite sense to the respective input (non- minimum phase behaviour). A small number of tests have been run with such processes. The results are:

LERNAS is able to handle friction and backlash, but here needs much more training than for smooth, continuous processes. So it is advisable, to pre- train and pre- tune (with respect to its free parameters) LERNAS on the basis of recorded process data from different process situations before it is used for on- line control.

Non- minimum phase processes cannot be handled by LERNAS in its present form. Possibly some higher monitoring level is necessary, which detects the unusual behaviour and corrects the plant inputs accordingly.

Dead- times are not a problem in principle. They do mean, however, that e.g. the predictive process model inputs are not adequately described by $[\underline{y}(k), \underline{y}(k-1) ...; \underline{u}(k), \underline{u}(k-1) ...]$ but by $[\underline{y}(k), \underline{y}(k-1) ...; \underline{u}(k-d), \underline{u}(k-d-1) ...]$, if a unique time delay $d \cdot T$ exists. No effort has been made so far to find an automatic selection procedure fixing the amount of history to be included into the situation-describing predictive process model input. Therefore dead times have to be adequately handled by the user of LERNAS himself in its current version.

III.2. Hierarchies and alternative structures

III.2.1. Structures for and preliminary experiences with LERNAS hierarchies

A very crude discussion as to why hierarchical structures of learning systems may be of interest and what problems arise in connection with hierarchies has been presented already at the end of section I.6.3. We shall take up here only a few further arguments and then concentrate on preliminary results.

From the point of view of technical processes, hierarchical control considered as a combination of specialization with respect to certain sub- tasks and of coordination as the tool to obtain the overall desired result, has the following advantages:

• sub- units of the general system are of lower complexity and can therefore individually be analyzed and controlled more easily.

- splitting the system into a number of individually controlled sub-units allows parallel working of the sub-units.

- the respective control-units can be brought nearer to the process (especially important for distributed processes) and malfunctions of one of the control-units may then be of small influence on the overall performance.

- changes in just one sub-unit are handled in general faster and more flexibly by a dedicated sub-controller than by an overall system controller.

Since technical processes are very often built up out of subprocesses directly, hierarchical control may fit very naturally into the overall system structure. Due to this fact a large amount of literature has evolved on the topic of decentralized and hierarchically integrated control together with the distinction between a number of general structures of plants and adequate hierarchical control schemes. This cannot be reviewed here but a basic discussion will be found in /15/. However, fig. 38 shows, as an example of the respective structures, the "multilevel system", which we are going to use after some remarks on hierarchies in biological systems.

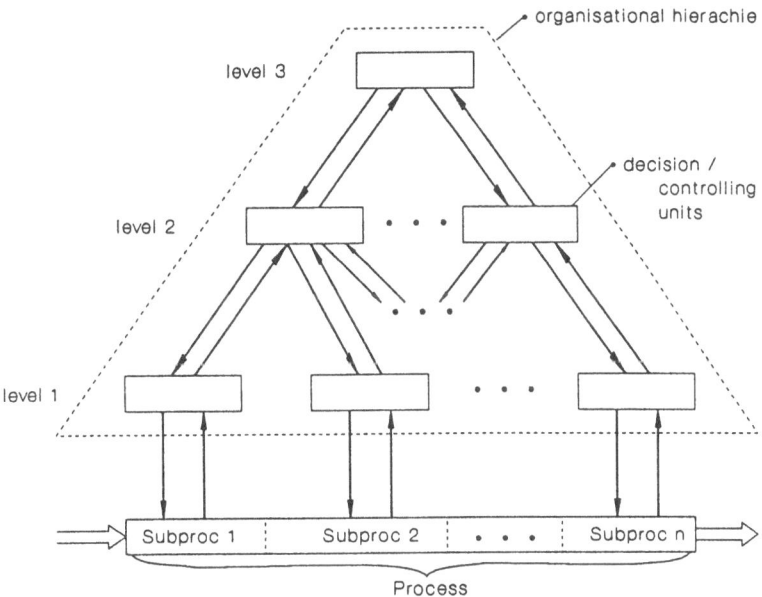

Fig. 38.: "Multilevel system" for a decentralized (in subprocesses divisible) process (example from possible hierarchical system structures).

With respect to biological systems there are a number of reasons to believe that their neuronal structures are built up as hierarchies, of which the lower levels concentrate on control functions and the higher levels progressively on coordinative functions. A very simple example (for which the frog would be an analogue biological system) shows the necessity in certain cases: The legs of a jumping jack can only move together. If one wants them to move separately, one has to cut the connection and build up a separate controller for each leg with a coordinating controller at a hierarchically higher level to restore the possibility of coordinated movements. Actually one can find such an evolution in the neuronal tissue of certain animals. In a more complex sense a multilevel hierarchy exists in the extra pyramidal motor system (see fig. I. 13) and it can be speculated, that hierarchical organization is also basic for general abstraction and thinking (see /16/).

Now, if hierarchies are important for technical processes as well as biological systems, it seems to be of interest to look at some basic questions of cooperation in learning systems. Taking into account the multilevel system structure of fig. 38 and the leg-coordination example, one is tempted to study the version shown in fig. 39b as a first step of such research.

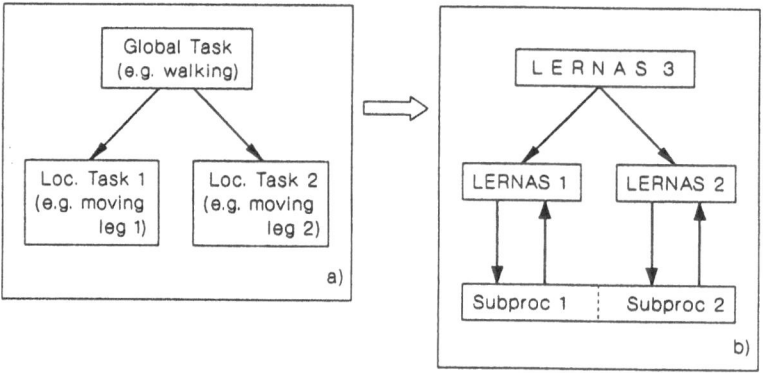

Fig. 39:   a) example, how multilevel systems can be motivated in biological systems,
             b) respective structure with LERNAS elements.

What problems have to be analyzed now for such structures?

Actually, four questions have been addressed in a first approach (see /15/ and/or /17/):

I.   What is a meaningful intervention from the coordinator onto the lower level systems?

II.   Is parallel learning in both levels possible or does a meaningful learning strategy require that the control of subtasks has to be learned initially before the coordination is learned (bottom up learning)?

III.   Normally, lower levels take care of short term requirements and the upper level of long term strategies. Is that necessary or what happens, if the upper level works on nearly the same time horizon as the lower levels?

IV.   Furthermore the upper level may look after other goals than the lower levels, e.g. minimization of overall energy consumption since the lower levels try to suppress disturbance effects. Can such different strategies work without oscillations or destabilization of the system?

Question I can be discussed by some general arguments, but as for questions II-IV up to now only indications of possible answers can be given based on simulation results.

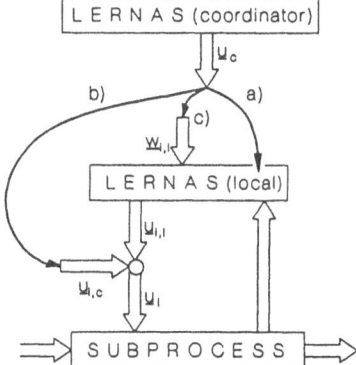

Fig. 40: Possible coordinator intervention schemes. For details, see text.

Fig. 40 shows three possible coordinator intervention schemes.

In case a) an intervention into the structure or the parameters of the sublevel (=local) controllers is meant. Since associative mappings like AMS have no parameters of major importance for the behaviour of the controller - as would be the case with parametrized linear or non-linear differential equations describing a conventional controller - this does not make sense for the controller, built up in LERNAS. However, one could consider the possibility of changing parameters or even elements (i.e. structural terms) of the performance criterion in LERNAS, which is responsible for the shaping of the controller. But this requires it to learn anew, which in general takes too much time.

In case b) a distribution of work load regarding control commands is meant, the idea being that the coordinator gives control inputs to hold the required long range mean value constant. The local controllers then just take care of fast dynamic fluctuations. However, this has the disadvantage that the control actions of the upper level have to be included in the inputs to the local storage devices, since otherwise the process appears to be highly time varying for the local controllers, which is difficult to handle for LERNAS.

So case c) seems to be the best solution. In this case the coordinators command the set points of the local controllers, thus generating local subgoals for the lower level controllers. Since this requires no input space extension for the local controllers and is in full agreement with the working conditions of simple LERNAS loops, it is a meaningful and effective approach. Also it is easily extendable to general hierarchies as shown in fig. 38.

Fig. 41 displays in detail the structure thus built up. The control strategy of Fig. I.5. in accordance with fig. 1-3 is split into the controller and active learning. Locally generalizing memories are the controller memories C1 - C3 and the model memories M1 - M3.

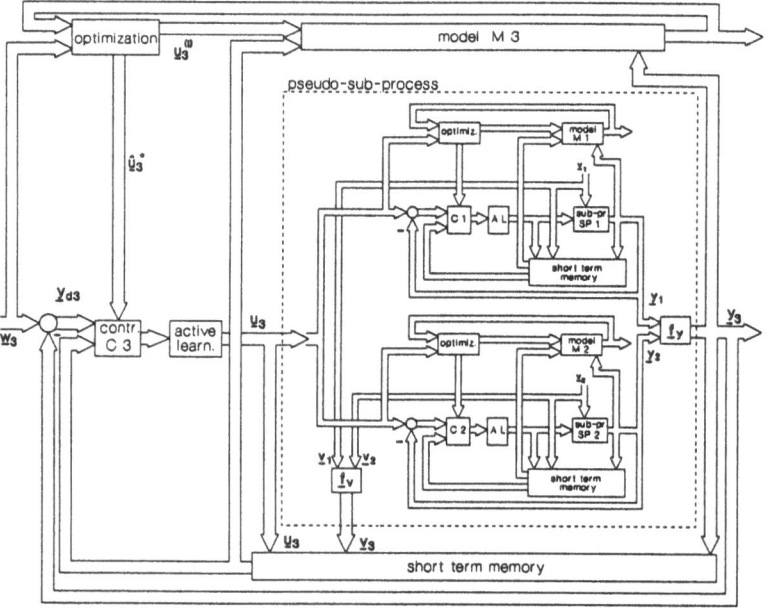

Fig. 41: Two level control hierarchy with coordination strategy c): OP = optimization, C = controller, AL = active learning, M = predictive process model, STM = short term memory, SP = subprocess. The subprocess for the coordinator (LERNAS 3) is shown by the dashed line.

In fig. 41 the elements are explicitly characterized for the upper level only. The whole lower level (compare fig. 39b) is considered by the coordinator as a single pseudo-process to be controlled.

To answer questions II and III a very simple non-linear process was used as shown in fig. 42b. LERNAS implies here - as shown in fig. 42a - the learning control loop without the unknown environment/process, i.e. the learning structure built around the process.

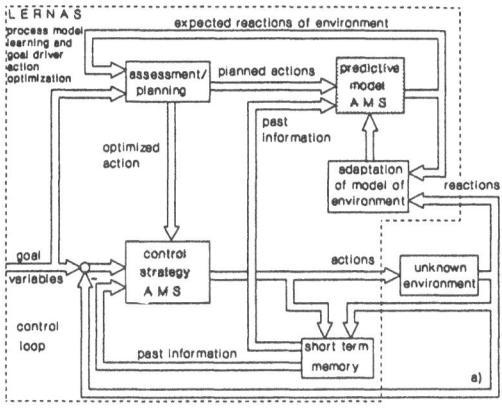

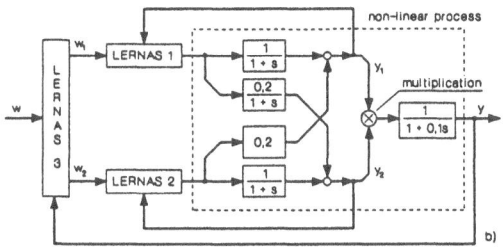

Fig. 42: a) LERNAS - definition as appropriate for hierarchical system description = learning structure without the already existing unknown environment/process. b) used simple non-linear process for investigation of questions II and III embedded in a structure according to fig. 41: For LERNAS 1, 2 the elements $1/(1+s)$ are the unknown subprocesses SP1, SP2, the crossfeeding being an unknown disturbance and the following multiplication with $1/(1+0,1s)$ being without influence. For LERNAS 3 everything between w_1, w_2 and y is the unknown process.

LERNAS 1 and LERNAS 2 consider the elements $1/(1+s)$ as their unknown subprocesses, the couplings being treated as unmeasured disturbances and the multiplication with the following element $1/(1+0,1s)$ being considered as not contained in the respective measurements. Actually the most simple LERNAS approach has been used here with $(i=1,2)$.

(51) $M_P : [y_i(k), u_i(k)] \rightarrow y_i(k+1)$
         $M_C : [y_{id}(k), y_i(k)] \rightarrow u(k)$
         $J(k) = |y_i(k+1) - w_i(k+1)| \equiv |y_{id}(k+1)| \rightarrow$ Min.

LERNAS 3 having LERNAS 1, LERNAS 2 and the whole nonlinear process as its respective unknown process is set up equally simply, however, with weighting of the command values for LERNAS 1, LERNAS 2 explicitly in its performance criterion:

(52) $\quad M_P : [y(k), u_1(k) \equiv w_1(k), u_2(k) \equiv w_2(k)] \longrightarrow y(k+1)$
$\quad\quad\quad M_C : [y_d(k), y(k)] \longrightarrow u(k)$
$\quad\quad\quad J(k) = 90 \, |y(k+1) - w(k+1)| + 5 \, |w_1(k)| + 5 \, |w_2(k)| \longrightarrow \text{Min}$

To discuss the question of whether the lower levels have to be trained first before the upper level (coordinator) can learn effectively - question II - as a first step the LERNAS 1, LERNAS 2 self-organizing controllers have been replaced by fixed PI-controllers of the following kind:

(53) $\quad\quad\quad u(k) = 0,4 \, [w(k) - y(k)] + u(k-1)$

With a sampling time of $T = 0,5$ sec this is equivalent to a PI-controller with a gain of $K = 0,4$ and an integration time constant of $T_I = 0,5$ sec.

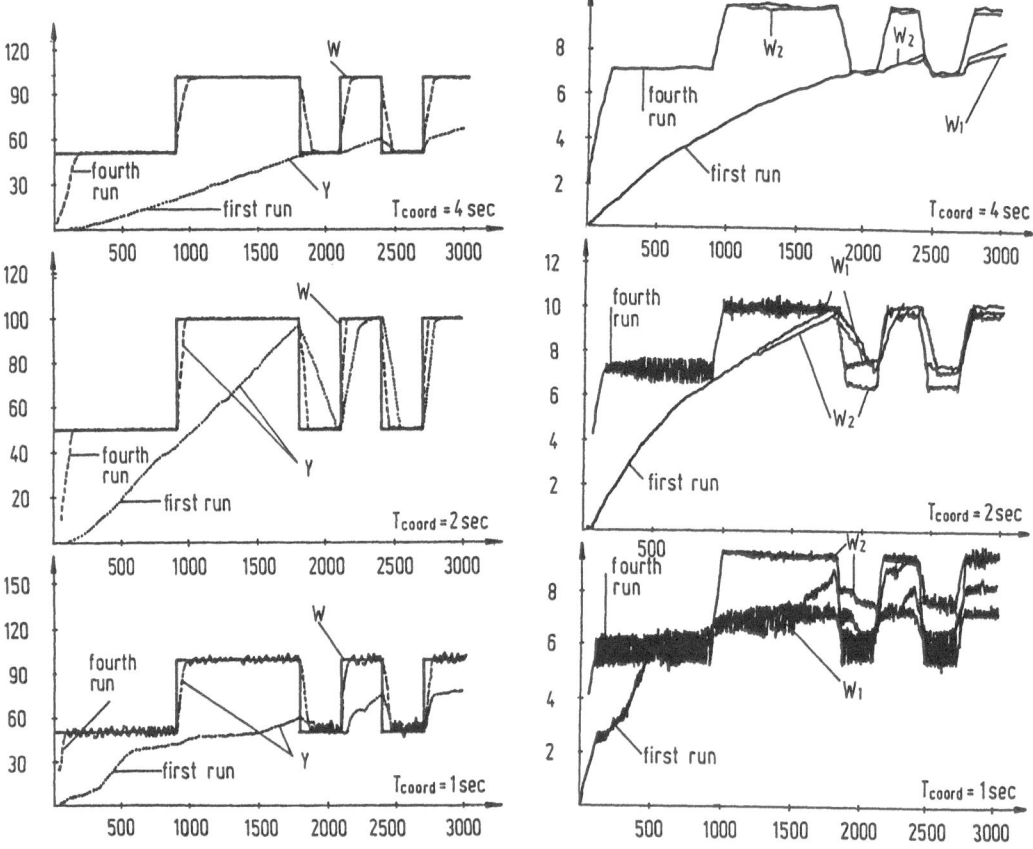

Fig. 43: Effect of changing the relationship of sampling times on subprocess control and coordination level. For the process from fig. 42 and an initiation of the trained situation at subprocess level by using there constant PI-controllers with a sampling rate of $T = 0,5$ sec, the learning behaviour regarding overall control (left hand figures) and resulting command inputs w_1, w_2 to the sublevel control loops (right hand figures) are displayed for the sampling time ratios 8, 4, 2. One sees, that learning of overall control is improved at first with a smaller ratio, but then deteriorated again as the ratio is going down further.

In a preliminary study the relationship between sampling times at the lower level and at the coordinator level has been looked at. Fig. 43 shows the result for the first run with empty coordinator memories and also for the fourth run (third repetition) with relationships of 8, 4 and 2 (coordinator sampling times of 4 sec, 2 sec and 1 sec). For the longest sampling time of the coordinator, the coordinator needs a relatively long period to learn how to control the process; however, in the fourth run the command values w_1, w_2 are specified in a way so as to obtain optimal overall control. With T_{coord} = 2 sec the learning of the coordinator is greatly improved but now certain problems in specifying w_1, w_2 arise. With T_{coord} = 1 sec. the problem of learning the overall control task and of coordinating the subcontrollers seems to become confused inhibiting the learning of overall control: At least to achieve this goal takes much longer again, than for T_{coord} = 2 sec. The results of fig. 43 were used to look at sequential and parallel learning, using the sampling periods T = 0,5 sec and T_{coord} = 2 sec in all test runs. One obtains the results displayed in fig. 44. In diagram a) the result of T_{coord} = 2 sec from fig. 43 is repeated, sequential learning being imitated by fully adjusted local controllers, here PI-controllers. In diagram b) from fig. 44 now LERNAS 1, LERNAS 2 and LERNAS 3 are used, starting with empty memories. One sees, that - at least in this very simple test case - parallel learning is possible and nearly as effective as sequential (bottom up) learning.

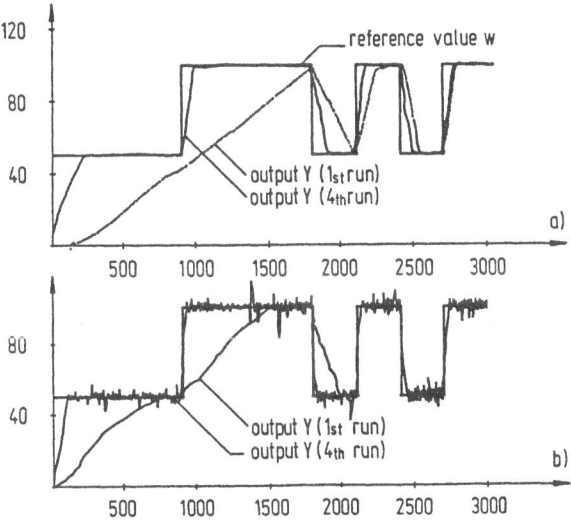

Fig. 44: Comparison of sequential and parallel learning. In diagram a) the controllers of LERNAS 1, LERNAS 2 are substituted by constant PI-controllers, imitating a fully trained sublevel status. In diagram b) all LERNAS loops 1 - 3 start with empty memories, learning by this in parallel (T = 0,5 sec, T_{coord} = 2 sec.)

Question III, regarding the optimal relationship of relative time horizons of upper and lower levels, can be answered for bottom up learning by looking at the cases of $T_{coord.}$ = 2 sec and $T_{coord.}$ = 4 sec in fig. 43. But is the situation the same or different for parallel learning? Fig. 45 shows the respective results for parallel learning and $T_{coord.}$ = 4 sec, the results for $T_{coord.}$ = 2 sec having been displayed already in fig. 44. One sees the same trends as for bottom up learning: smaller sampling rates allow the coordinator to get more information about the pseudo-sub-process and by this means the global goal is reached faster. However, larger sampling rates lead to a better overall performance when the control goal is reached, since a higher amount of averaging regarding information about the pseudo-sub-process takes place.

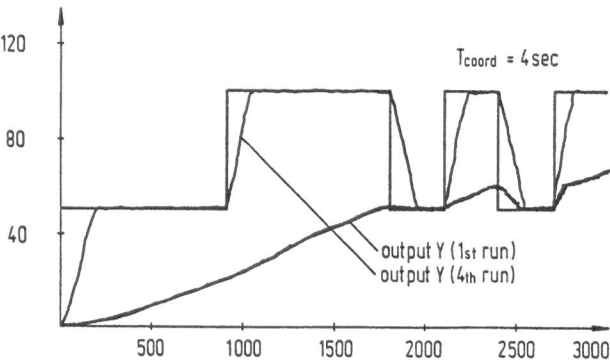

Fig. 45: Parallel learning with T = 0,5 sec, $T_{coord.}$ = 4 sec.

Finally regarding question IV, the behaviour with different control goals (performance criteria) on the lower and the upper level of the learning control loop, a very great variety of possible task formulations exist, and the answer may depend on the chosen goal combination. Some investigations have been performed in /17/ and, actually, no oscillating and/or destabilizing behaviour has been observed with respect to the overall control loop. The example considered in /17/ was a simulated two stage pH-neutralization process as shown in fig. 46, each stage considered to be a subprocess for which the higher level LERNAS 3 - coordinator is setting the reference values (see fig. 41). Each subprocess corresponds to a neutralization process as described in section III.1.4; therefore further details will not be discussed here. Since on the subprocess level the performance criteria consider the control deviations y_d, the coordinator performance criterion is selected to minimize the overall neutralization effort by taking into account a weighted sum of the variances of the reactor inputs u_1, u_2:

$$(54) \qquad J_{coord.}(k) = r_1 \left[\frac{1}{n} \sum_{i=0}^{n-1} u_1^2(k-i) - \left\{ \frac{1}{n} \sum_{i=0}^{n-1} u_1(k-i) \right\}^2 \right] +$$

$$+ r_2 \left[\frac{1}{n} \sum_{i=0}^{n-1} u_2^2(k-i) - \left\{ \frac{1}{n} \sum_{i=0}^{n-1} u_2(k-i) \right\}^2 \right]$$

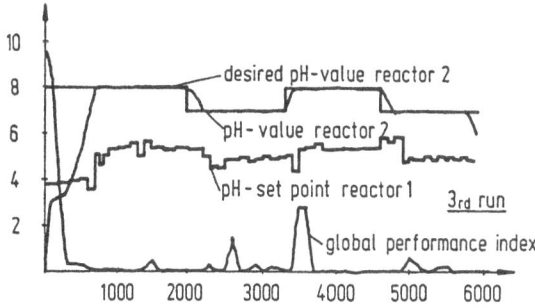

Fig. 46: Two stage pH-neutralization process.

Fig. 47 shows the result after a second repetition (third run) for a $T_{coord.}$ to T_{basic} sampling relationship of 50. The set point for reactor 2 is the overall required reference value w. Displayed are the obtained pH-value = y and the set point of reactor 1 as put forward by the coordinator. One sees that the coordinator distributes the work load between the two neutralization stages by moving the set point for reactor 1 up and down in accordance with the requirements stemming from the reference value w. No oscillations and/or destabilization effects take place. The learning is, by the way, not completed, as one can see from the peaks in the performance index curve being caused by active learning.

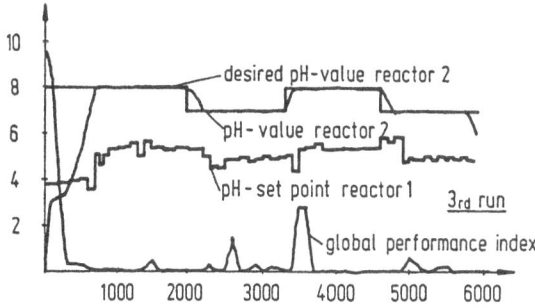

Fig. 47: Global performance criterion value (54) and coordinator set point specification as well as system answer for the two stage pH-neutralization prozess of fig. 46, situation after three learning steps, $T_{coord.}/T = 50$.

In conclusion one can state that hierarchies of learning control loops seem to be possible. By using higher levels of the hierarchy to create appropriate reference values for the lower levels the locally generalizing memories necessary to build up predictive process models and advantageous controllers may be more simple in the sense of fewer necessary inputs for describing the actual situation. A consequence is the reduction of necessary dimension and use of fewer memory cells (on account of more precise estimation of ranges for the inputs), than in the case of applying one LERNAS system to the whole complex unknown structure. It is easier to exploit structural knowledge about the process. However, real applications have not so far been investigated and further research still has to be performed in this area, especially with respect to the question, of whether some self-organizing adaptation to process structures by self-forming hierarchies is also possible.

III.2.2. Changes and alternatives to LERNAS

III.2.2.1. Prestructuring of the locally generalizing memories

Up to now we have mainly discussed what we can achieve with a learning control loop using as small an amount of information about the process and about control theory as possible. The applied knowledge was a rough idea about process behaviour for the selection of the sampling time, for the selection of necessary history to be included in situation description and for selection of input and output limits to estimate the necessary memory sizes. For complicated processes a certain off-line experimentation with respect to these values based on measured and stored process data may be helpful or even necessary to achieve good overall performance.

Although the simple engineering knowledge above mentioned proved to be sufficient, additional knowledge about the process and some adequate controller can, of course, be explored to simplify or improve the learning control loop.

For the process we often know a set of differential equations, describing its structure fairly accurately, but being uncertain about the respective parameters, which may depend on the current process situation. Since a differential equation gives the amount of change of the process describing values (state variables) and thus a prediction of their behaviour, the learning of the process model may be substituted by learning of the parameter values for different situations into the locally generalizing memory used for process behaviour predictions and thus the necessary size of the memory may be substantially reduced, however, paid for by some additional computational load.

An example of this effect has already been mentioned in section I.6.1. For a robot manipulator the dynamic equations are (see e.g. /18/):

$$(55) \qquad \ddot{q} = H^{-1}(q) \, [\underline{\tau} - D(q,\dot{q})] \equiv \underline{\varphi}(q,\dot{q},\underline{\tau})$$

where $\underline{q}$ is the vector of joint variables, $\underline{\tau}$ the vector of forces and/or moments applied in each of the one degree of freedom joints, H the matrix of inertia and D a summation of Coriolis and gravitational forces. For a six-degree of freedom robot one gets without considering the structure of equation (55) a memory with eighteen inputs. However, one can write equation (55) with the knowledge of its structure also:

$$
\begin{aligned}
(56) \qquad \ddot{\underline{q}} &= H^{-1}(\underline{q})\,\underline{\tau} - H^{-1}(\underline{q})\,D(\underline{q},\dot{\underline{q}}) \\
&= \varphi_1(\underline{q},\underline{\tau}) + \varphi_2(\underline{q},\dot{\underline{q}}) \\
&= \underline{\Phi}(\underline{q})\,\underline{\tau} + \varphi_2(\underline{q},\dot{\underline{q}})
\end{aligned}
$$

that means one can store the changes of $\dot{\underline{q}}$ (the $\dot{\underline{q}}$-prediction) also, either by two tables with 12 inputs each and just one summation of their outputs or by one table with 12 inputs and one table with six inputs, needing now some multiplications together with the summation. We shall not go into any details of application of these ideas, but the reader can find some relevant remarks, connected however with the problem of inverse kinematics in /19/ (see also I.6.1.).

Turning now to controller knowledge we remember that continuous, e.g. constant disturbances pose a problem for learning control loops in so far as they involve a change of the process as seen by the learning loop, which implies re-learning. Actually, the learning control loop looks at part time constant disturbances like an adaptive controller.

However, we know from control theory that for handling disturbances of known structure we do not need adaptation, but only some inner model of the disturbance structure - see e.g. /20/ or /21/ -. Especially for constant disturbances the inclusion of an integrating element into the controller suffices. This leads naturally to the idea, of structuring the controller not by learning some general non-linear mapping but rather by learning and locally generalizing the parameters for an adequate controller depending on the particular circumstances (e.g. a PI-controller).

But now a number of particularities has to be considered.

Firstly, to judge the performance of a controller of a certain kind one has to take into account the overall closed loop behaviour. That means one cannot use just a one step ahead prediction. However, long prediction horizons run into difficulties relatively often, since one cannot expect that they are always leading to trained area, especially at the beginning of learning about the plant. Two possibilities exist:

- either one works for the process as well with some structure - e.g. the frequently satisfactory $PIDT_1$-T_t structure - which allows a direct selection of advantageous controller parameters for each learned and/or interpolated parameter set

- or one prescribes transition trajectories for set point changes by shaping the desired output = w-changes through an appropriate prefilter: here one uses in the multi-step prediction the known w(k+1), w(k+2) series to calculate, with the help of the performance criterion, favourable controller parameters.

Up to now only the second one of these ideas has been studied (cf. Ersü and Wienand in /22/).

A second question, which needs to be addressed for prestructured controllers, which imitate linear approaches like the PI-controller, is the stability problem. Here a strong input limitation $\underline{u}_{max} \leq \underline{u} \leq u_{min}$ no longer exists (although, surely, some limitations will be present in practice) and therefore the plant may be shifted unwittingly into areas of high excitation, a dangerous situation which may occur most easily at the beginning of the learning process, when only little process knowledge exists and consequently appropriate updating of the controller parameters may take place too late. Ersü and Wienand propose as a solution to this problem the inclusion of an additional constant PI-controller, from which no performance optimization is demanded but only an ability to stabilize the plant. In detail, the control loop works as follows:

- if sufficient plant knowledge is available, an optimal parameter set $\underline{p}_c^*(k)$ is computed for the controller in the process situation considered. The prediction length, l, for this calculation is variable between some limits $l_{min} \leq l \leq l_{max}$, with l_{min} being the shortest prediction length (transition trajectory length) which allows a fairly accurate selection of the parameters and l_{max} being an estimate of the longest prediction, which may give a quite accurate forecast of the process behaviour. $\underline{p}_c^*(k)$ is used in the same way as $\hat{\underline{u}}^*(k)$ in LERNAS with an unstructured controller memory as something to be stored away and to be applied as a starting point for the optimization procedure and also for giving an input $\hat{\underline{u}}^*$ to the active learning block in case of simulations (for the appropriate sequencing in real-time applications, see section III.3.1). Active learning changes the output of the optimized PI-controller $\hat{\underline{u}}^*$ again in such a way that further knowledge about the process behaviour is obtained.

- if the plant knowledge turns out to be insufficient for the minimal prediction horizon l_{min}, normally the parameter set $\underline{p}_c^*(k-1)$ is used for the controller (so that $\underline{p}_c^*(k)$ is kept in addition to its storage into the locally generalizing controller parameter memory for just one sampling step in a further short term memory). This is possible in contrast to the case of normal LERNAS in which only $\hat{\underline{u}}^*(k)$ is adapted directly to the given situation, since the PI-controller is with any parameter set a global controller and if the behaviour of the plant has not changed drastically during this single sampling step due to the non-linear controller, $\underline{p}_c^*(k-1)$ will be in general also a good estimate for $\underline{p}_c^*(k)$.

However, a drastic change of the plant behaviour is not impossible and through such a change furthermore the closed loop stability may be endangered. Therefore a supervisory level was introduced in addition, monitoring whether the control deviation is liable to drift away eventually more than a preset limit. In that case the parameter set of the demanded overall stabilizing PI- controller is used instead of $\underset{\sim}{p}_c^*(k-1)$.

The parameter set of the overall stabilizing PI controller is used also, if neither at time kT nor at time (k- 1)T an optimization was possible, which means nearly always at the beginning of the process knowledge collection.

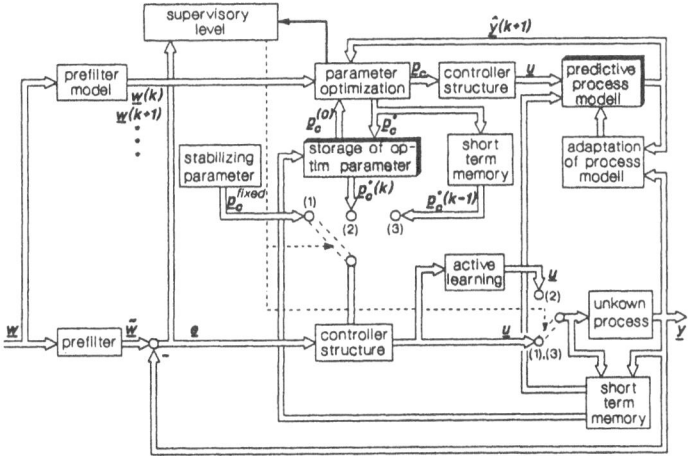

<u>Fig. 48.:</u> Learning control loop with an optimized (trained) prestructured controller and a stabilizing back up controller. The inputs are shaped by a prefilter and the input changes predicted for the optimization by a model of the prefilter. A supervisory level detects how far optimization was possible at the actual time kT or eventually at the preceding time (k- 1) T and/or if safety limits for the control deviations e_i are possibly exceeded. It selects the optimal parameters $\underset{\sim}{p}_c^*(k)$, if they do not exist $\underset{\sim}{p}_c^*(k-1)$ and if these also do not exist or if the safety limits are violated the stabilizing parameters $\underset{\sim}{p}_c^{fixed}$. With $\underset{\sim}{p}_c^*(k)$ active learning is used to extend process knowledge, otherwise the process will be lead into unknown area, anyhow. All further differences to fig. I.5 are due to the fact, that now controller parameters are stored instead of a controller strategy.

The advantages of the control loop scheme just discussed, shown in fig. 48 as a block diagram, are:

• the structuring of the controller reduces the necessary amount of locally generalizing memory, because each parameter set can be used for a greater area than a control value $\underset{\sim}{u}$ and therefore it is not necessary to distinguish as many process situations as in the latter case.

- for the same reason learning takes place more quickly: A parameter set once learned can be used in more situations than a learned optimal control value and by this less different parameter sets have to be learned than otherwise control values.

- properties due to the controller structure, like constant disturbances rejection by the PI- controller, shall be present in the loop.

However, as a disadvantage one has to expect, that the prescribed controller structure in accordance with its more global reaction has also some smoothing effect and because of this difficulties with very strong non- linearities.

III.2.2.2. Results with a prestructured controller

The structure of fig. 48 has been implemented and investigated by using the pH control problem described in section III.1.5.1. as the process example. For the controller the PI-structure

(57) $u(k) = u(k-1) + q_0[w(k)- y(k)] + q_1[w(k-1)- y(k-1)]$

with the constraints $q_0 > 0$, $q_1 > -q_0$ was used. The fixed values for the back up controller were $q_0 = 0,5$ and $q_1 = -0,25$, the overall sampling time being $T = 2$ sec. The prediction length for generating learned q_0, q_1- values has been chosen automatically - as mentioned in the last section - between $l_{min} = 4$ and $l_{max} = 10$, and for the desired transition trajectory a second order system described by the Laplace transfer function $G = \dfrac{1}{1+2\cdot 20s +20^2 s^2}$ (no overshoot) has been used.

At first, the problem of reaching parameter values to control some constant pH- value, like pH=9, pH = 8 and pH = 7 was studied. Since it is not a problem to keep pH equal to 9 fairly accurately, the steepness of the titration curve for pH=8 and pH=7 leads to difficulties. The reason is that near the saturation region (pH $\approx$ 10), the controller gain has to be relatively large to obtain any changes in the behaviour at all. For regions where the non- linear characteristic is steeper, the gain has to be reduced. Actually the control scheme acts accordingly (see fig. 49b). However, when the characteristic changes very rapidly, the smoothing by the linear PI- controller together with the multi- step prediction procedure is not cautious enough and one gets some relatively high overshoot beyond the demanded pH- values and consequently oscillations which can be still reduced for pH = 8 by learning but no longer for pH = 7 (see fig. 49 a). The behaviour is shown for the first run, but the situation is not improved much by further optimization/learning. Naturally, the results depend also on the chosen PI- back up controller, which is not able to handle pH = 7 satisfactorily, this being itself a very difficult problem for a fixed linear controller. Also, the idea of using a learning control loop is just that no extensive search for some very advantageous back- up controller should be necessary. In contrast to the learning control loop with a controller of PI- structure the original

LERNAS system without any controller structure is able to handle pH = 7 not only as shown in fig. 45 by coming from pH = 9 but also by going to pH = 7 directly from pH = 3 - see fig. 50. This is due to the improved possibilities of adaptation with the unstructured general mathematical mapping used in the normal LERNAS version.

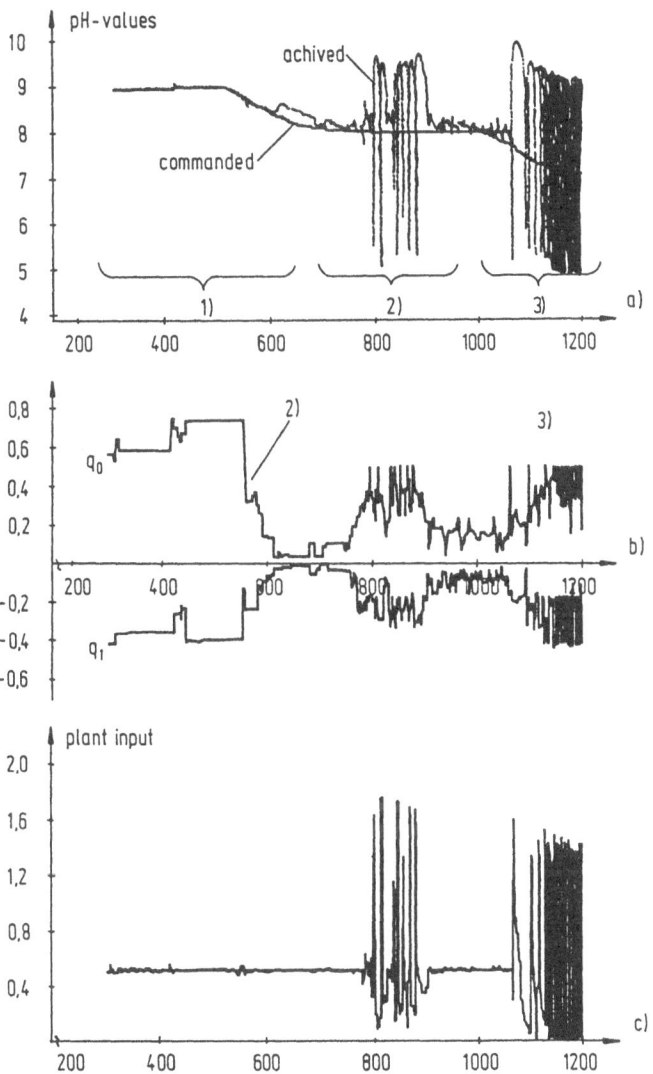

Fig. 49: pH- control with back up PI- controller (q_0 = 0,5, q_1 = - 0,25) and learning of situation dependent q_0, q_1 - values; structure of the learning control loop given in fig. 48. a) = achieved behaviour of plant output, c) respective plant inputs, b) = actual q_0, q_1-values, pH = 9 = 1) learned easily, pH = 8 = 2) oscillations, however, damped out, pH = 7 = 3) uncontrollable case; first run; but not to be improved much by further learning.

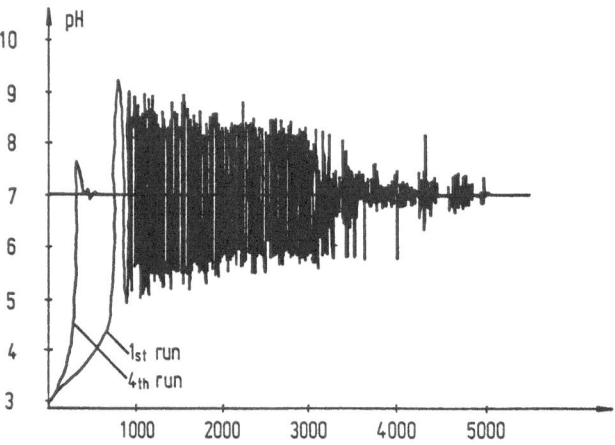

Fig. 50: Stabilization of pH- control for pH = 7 by LERNAS without PI- controller structure, first and fourth run.

By means of this result the correctness of the expectation at the end of the last section is proven, that the smoothing effect of a prescribed controller structure may lead to difficulties with very strong non- linearities, this being not the case without such a pre- structuring.

Now, some solution to handle pH control by LERNAS with PI- controller structure exists, too. The necessary measure is to give more information about the changes stemming from the steeper slope of the titration curve to the learning control loop. This can be achieved by reducing the quantization of the measured quantities as well as the sampling period. But due to noise especially a very fine quantization of measured quantities will not be possible in practice. However, some simulation of such an approach can be reached by stretching the titration curve with respect to its abscissa. Fig. 51 shows the respective characteristic used instead of the characteristic of fig. I.7.

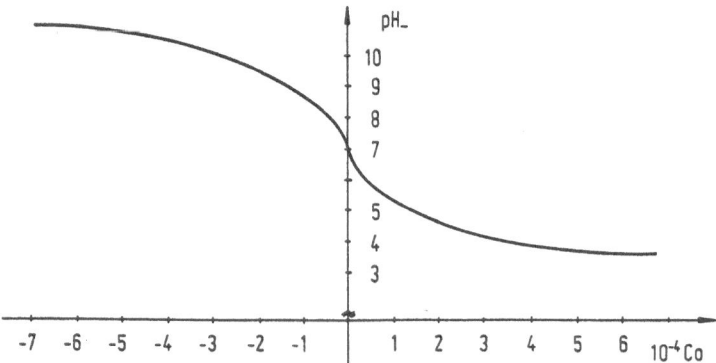

Fig. 51: Stretched titration characteristic to simulate high quantization of measured values.

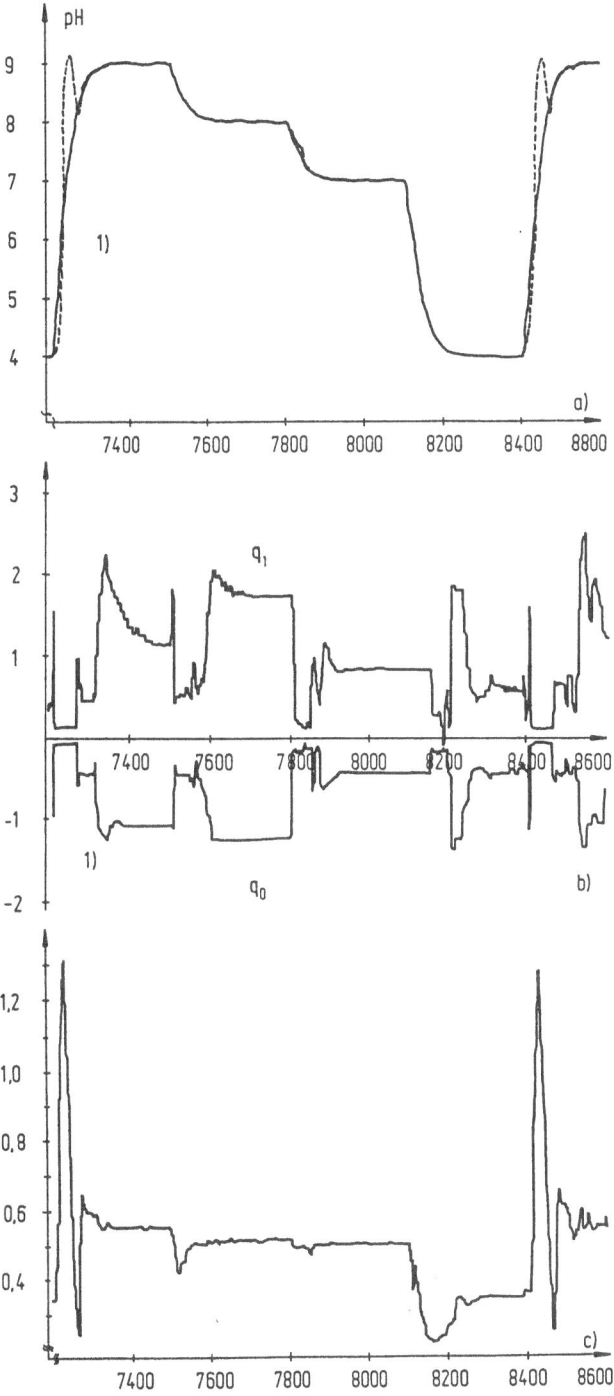

Fig. 52: pH-control with stretched titration curve from fig. 51, 4th run (third repetition): a) pH value control, b) learned q_o, q_1- values, c) plant input; 1) still not enough training/optimization.

The LERNAS system with PI-controller structure is now easily able to handle the plant for all possibilities of desired pH-values as shown in fig. 52. Here the results of the fourth run are displayed. Only some problems exist still when jumping from pH = 4 to pH = 9, which is due to the fact that in a big jump only a little information is collected for optimization, so that a higher number of such jumps would be necessary to obtain a fully satisfactory transition.

Having demonstrated earlier the problem of structured controllers with strong non-linearities, we can now exemplify the advantages of the PI-structure for suppression of disturbance effects. For this purpose we consider a trained control loop with the learning level shut off. The overall behaviour is demonstrated by going through a sequence of different pH-values and by making simultaniously, but not at the same time instants, large changes in the unmeasured disturbance inputs concentration c_i ($\approx$ 100 %, PT1 with T = 20 sec) and mass flow $\dot{m}_i$ ($\simeq$ 50 %, PT1 with T = 40 sec). Fig. 53 shows, that these strong disturbances cause only very small deviations from the desired value even in the very sensitive case of pH = 7. This has to be compared with the LERNAS behaviour without PI-controller structure displayed in fig. 11, where control difficulties arise even for disturbances of the order of 1 % for the less sensitive case of pH = 8 so long as the learning control loop is shut off.

In conclusion one can state, that structuring of the locally generalizing memories of LERNAS is on one hand reducing its adaptibility to the problems occuring but may on the other hand improve the behaviour of the control loop in the direction of the theoretical abilities of the considered structures and will reduce in any case the necessary amount of memory size and learning.

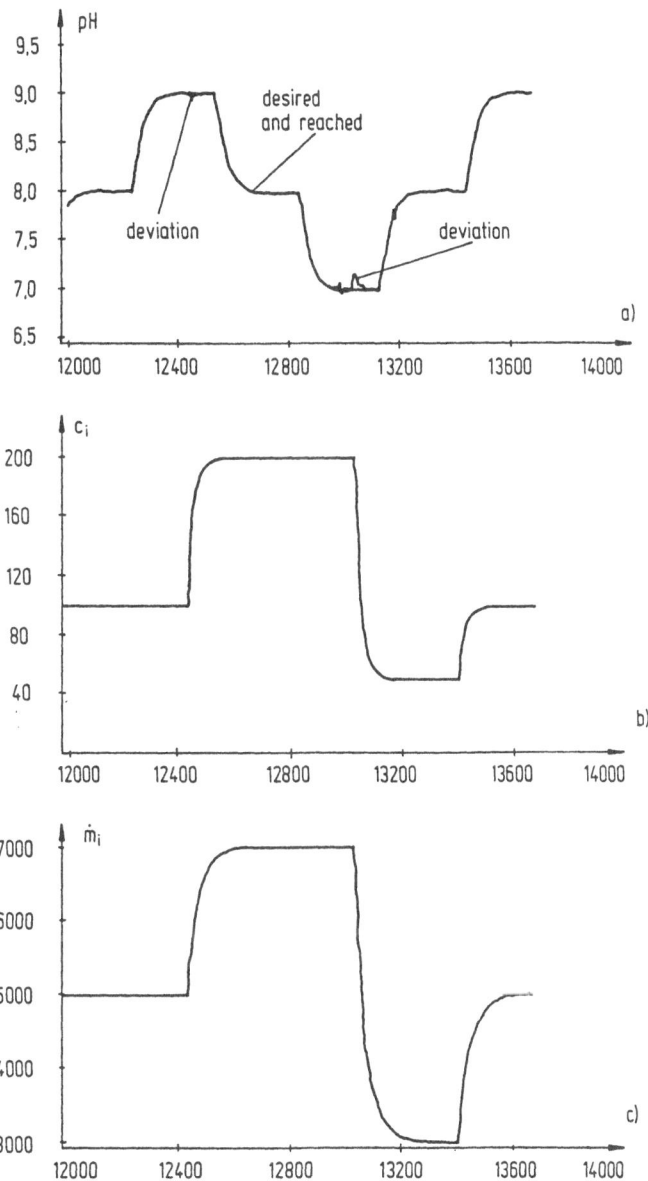

Fig. 53: Disturbance rejection of LERNAS with PI-controller structure: pH-neutralization for stretched titration curve, input concentration c_i and input mass flow changes $\dot{m}_i$ of ~ 100 % and/or 50 % for pH = 9 and the very sensitive case pH = 7; a) pH desired and reached values b) input concentration, not measured; c) input mass flow, not measured.

III.2.2.3. Simplifications

LERNAS works with an explicit process model. However, this seems not always to be the case with human beings. In very simple tasks as well as with very complex processes we seem not to learn process details but just to test and to remember, whether or not in a certain situation some action proves to be advantageous. This behaviour has been taken up and modelled by E. Ersü in a simplified LERNAS structure, called MINLERNAS (/23/). Fig. 54 shows the respective learning control loop.

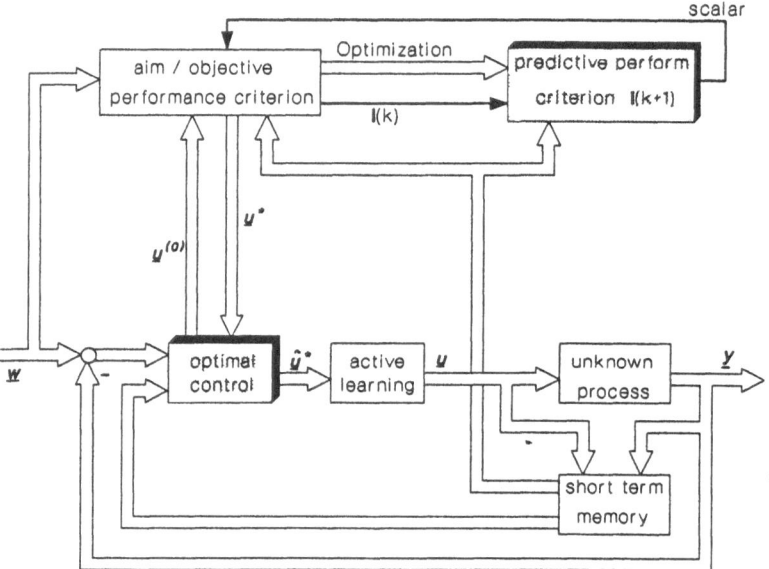

Fig. 54: Learning control loop with implicit process modelling only: Besides of the optimal control just a prediction of the performance criterion value for the respective situation is stored in a locally generalizing associative memory. The optimal control is calculated by looking for the best reachable performance criterion value in the next control step. The knowledge about the result of actions in eventual situations is again extended by active learning (compare with fig. I.5 regarding learning control loop with explicit process modelling - LERNAS-).

Instead of the predictive process model, the performance criterion J(k) is stored in a locally generalizing memory as a prediction of how far some action $\underline{u}(k)$ may improve the process control. $J(k) = L [\underline{y}(k+1), \underline{w}(k+1), \underline{u}(k)]$ due to (3) and one can approximate in this expression $\underline{w}(k+1)$ by $\underline{w}(k)$ as usual. However, $\underline{y}(k+1)$ is here no longer available from an explicit predictive model. But $\underline{y}(k+1)$ is - cf (1), (2) - a unique function from $\underline{y}(k), \underline{y}(k-1) ..., \underline{u}(k-1) ..., \underline{v}(k-1) ...$, so that J(k) may be learned and stored directly as a function of this history leading to the mapping:

(58) $[\underline{w}(k), \underline{y}(k), \underline{u}(k)] \rightarrow J(k).$

By using some optimization procedure one can now find with the help of (58) for trained situations the best $\underline{u}(k)$ to be applied nd by active learning one can extend the knowledge about the mapping (58) systematically. For a storage scheme one may choose any of the discussed locally generalizing memory systems, AMS, AMS with variable generalization, MIAS or even some other storage device, which is able to interpolate automatically between scattered data.

The major differences between LERNAS and MINLERNAS are:

- the amount of memory necessary for the learning loop is reduced for multi-input, multi-output plants in MINLERNAS, since the performance criterion used for control input selection is always a scalar.

- multi-step predictions are not possible with MINLERNAS, since due to the missing predictive process model the new situation reached after applying $\underline{u}(k)$ cannot be calculated.

MINLERNAS is, by the way, not the only possible macrostructure simplification. Fig. 55 shows some further complexity reduction, studied by W.T. Miller III and R.P. Hewes (/24/). Here no explicit performance criterion exists, only the general requirement for diminishing the control deviation is used for learning by alternating between training and control application cycles. Actually some basic linear controller is used to hold the loop output in certain limits independent of disturbances. So far this scheme is similar to the back up system of learned PI-control. In using now only every other sampling step for control application, the free sampling steps in between can be applied to build up some additional non-linear control input. Here one attempts to reduce the control deviation $\underline{w}(k) - \underline{y}(k)$ to zero by applying $\underline{u} + \Delta\underline{u}$ instead of $\underline{u}$ and learning in this way an appropriate $\Delta\underline{u}$. The cerebellar model articulation controller "CMAC" of Albus - see section II.1.2 - is used by Miller in some own software implementation to build up and store the stepwise learned mapping between $\Delta\underline{u}$ and $\underline{y}(k)$, $\underline{w}(k)$ in a locally generalizing fashion. As an application a robot control task has been taken and it was shown, that the control loop from fig. 55 works very satisfactorily.

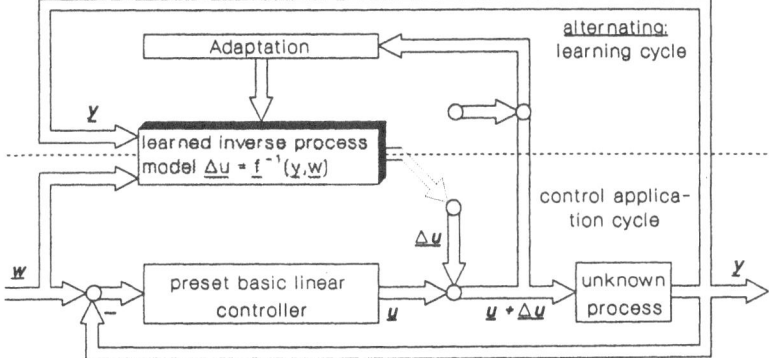

Fig. 55: Improvement of linear control by learning in an alternating fashion to the control application additional non-linear control features (W.T. Miller et.al. /25/).

In the context of the discussion given in this book, Miller's learning control loop can be consid red as an intermediate structure between MINLERNAS and the direct application of an off-line learned inverse process model as a precompensator for linearization and decoupling in an ordinary control loop (see section II.6). To place the process inversion outside the actual control loop and use it as an additional control input only is a usual approach which is best to avoid that differences between the process and an inverse process model can endanger the closed loop stability. As with the PI-structured controller a combination of control theory knowledge and the concept of learning is applied.

III.2.2.4. Results achieved with MINLERNAS

MINLERNAS has been applied as a first step to a simple PT1-process to study its fundamental behaviour and in a second step to the pH-neutralization problem in order to demonstrate its ability to handle real, highly nonlinear plants.

For the PT1-process a time constant of 16,7 sec. was chosen, leading to

$$(59) \qquad y(s) = \frac{1}{1 + 16,7 \ s} \cdot u(s)$$

and a sampling time of 10 sec. Fig. 56 shows the results achieved with a generalization of $r^* = 8$ for the performance criterion AMS-type memory and of $r^* = 4$ for the controller AMS-type memory. The repetition of the control task is reached here by jumping up and down between w = 5 and w = 0 instead of starting new runs from the same initial conditions. As performance criterion the absolute value of control deviation has been used together with a weighting of actuator (control input) activity. Fig. 56a) displays a good learning success for the first changes of the requested output, but only a fair overall control performance in the long run. Since this is not the case with LERNAS - respective results can be found in /27/ - this indicates - as may have been expected - that the implicit modelling of the process through the performance index is less effective than working with a direct predictive process model in the learning/planning level. A confirmation of this statement was given by the fact, that with MINLERNAS in this basic form it was not possible to control the pH-neutralization, which was easily performed by LERNAS.

175

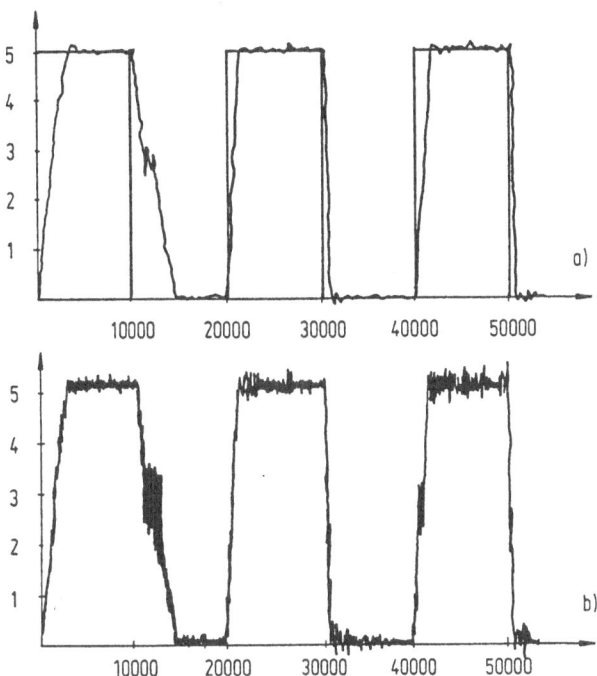

Fig. 56: Learning of MINLERNAS to control a PT1-process with $r^* = 8$ for the performance criterion AMS and $r^* = 4$ for the controller AMS. a) achieved outputs, b) respective control inputs (cf /23/ and/or /26/).

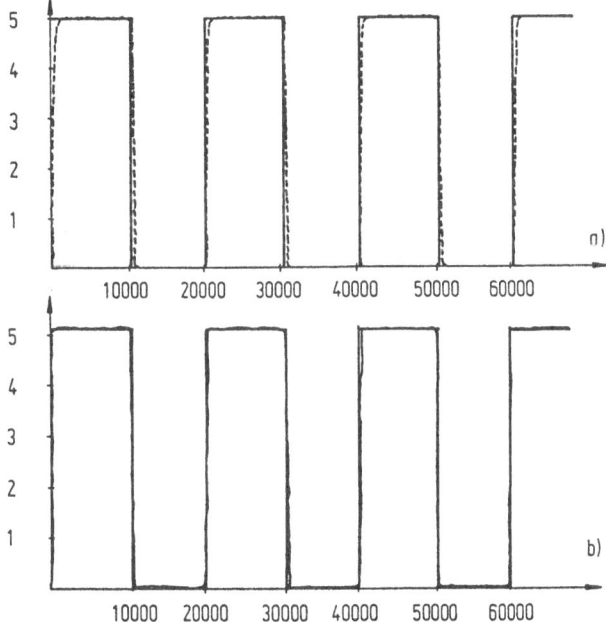

Fig. 57: Learning of MINLERNAS with variable generalization ($r^* = 16, 8, 4, 1$) for both the performance criterion and the controller memory. The process is the same as in fig. 56; a) gives the achieved output behaviour, b) the respective control inputs.

However, the MINLERNAS achievements can be very much improved by variable generalization and the application of an incremental controller, for which the actual output to the plant· is calculated by

$$(60) \qquad \underline{u}(k) = \underline{u}(k-1) + \underline{\Delta u}(k)$$

and where only $\underline{\Delta u}(k)$ is memorized. As variable generalization the parallel use of AMS- memories with different r^* values has been applied as explained in section II.5.2 and fig. II.20. Fig. 57 shows the respective results for the PT1- process (59).

One finds an excellent tracking behaviour as well as a very smooth control input activity. The variable generalization was used for the performance criterion memory as well as the controller memory with four different r^*-values: 16, 8, 4, 1. $r^* = 1$ is actually only a look up table with no local generalization. In fig. II.21 the general behaviour for the controller memory has been displayed already: At first mainly the broad generalization of $r^* = 16$ is used, switching then to an additional use of the $r^* = 8$ and $r^* = 4$ generalisation and making use finally of the look up table ($r^* = 1$) also.

Naturally, these very good results led to the application of MINLERNAS with variable generaliza- tion to the neutralization process, too. Fig. 58a) shows the results of the initial run, starting from pH = 3 with empty memories and requesting at first to master pH = 9, then pH = 8 and finally pH = 7, a task which could not be handled by LERNAS with a PI- structured controller. MINLERNAS with variable generalization is able to control the pH- neutralization from the beginning for all three set- points. Since for comparison the same problem is solved with LERNAS - see fig. 58b) -, one can compare the behaviour of the two different learning control loops. Actually, the behaviour is very similar, the main difference being, that LERNAS even with fixed generalization achieves a good control much faster than MINLERNAS with variable generalization. Taking into account the different time scales in fig. 58a), b) one sees especially, that for the most difficult case pH = 7 one needs with an implicit process model (MINLERNAS) twice the time to hold the set point without further oscillations than with an explicit process model (LERNAS).

However, the major conclusion of this chapter has to be that MINLERNAS with variable generali- zation and an incremental controller is a real alternative to LERNAS for handling even very nonlinear unknown processes, which means that one does not need an explicit process model for this purpose.

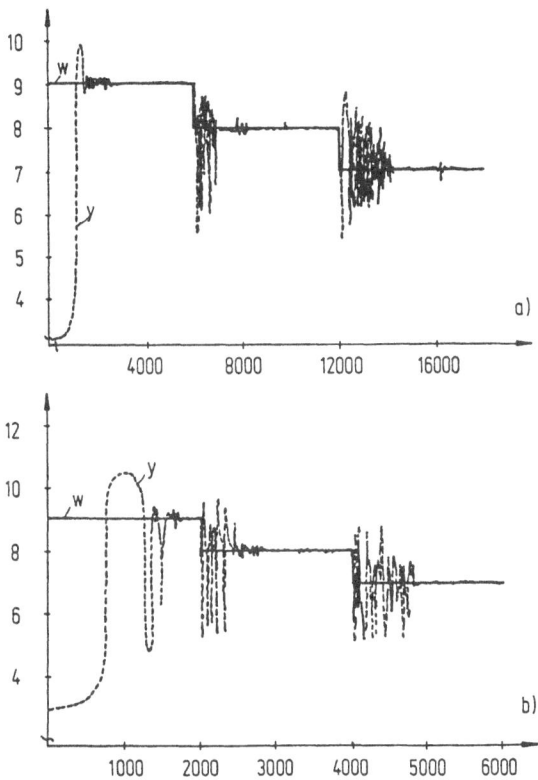

Fig. 58: Handling of pH control for three different set points by MINLERNAS with variable generalisation (a)) and by LERNAS (b)). Both learning control loops can achieve the keeping of the required set-points even in this shown first run. However - compare the different time scales - LERNAS reaches a good control behaviour much faster than MINLERNAS.

III.2.2.5. Results achieved with Miller's Learning Control loop

Fig. 59 gives a sketch of the used implementation of Miller's learning control loop. The updating of the learned control improvement is done according to

$$(61) \qquad u_{AMS}^{new}(k) = u_{AMS}^{old}(k) + K_R \beta [w(k) - y(k+1)]$$

In this formula represents K_R the gain of the conventional controller with the intention to keep u_{AMS} in the same order as the output u_R from the conventional controller. β, a factor between zero and one determines, to what extent the control difference (w - y) shall be considered in the modification of the learned control u_{AMS}. It is a tuning parameter of the Miller approach. That (w(k) - y(k+1)) is used in (61) and not w(k) - y(k) and/or w(k+1) - y(k+1) has reasons of practical program shaping only and is of no real importance.

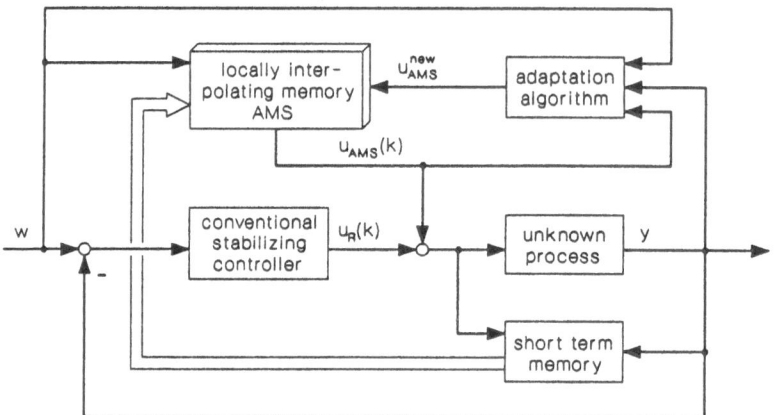

Fig. 59: Single-input, single output implementation of Miller's learning control loop; general layout, see fig. 55

For the selection of the amount of history, one has to take into account fig. 60. That means, a high order conventional controller may necessitate to take more history into account than in other learning control loop approaches being necessary.

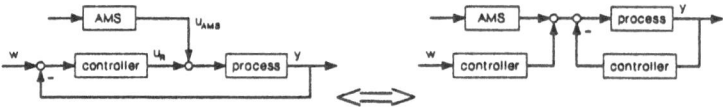

Fig. 60: Equivalent structures, showing, that for the selection of necessary history in the Miller control loop not the process itself has to be considered, but a pseudo-process, consisting of the closed loop around the process with the chosen conventional controller.

Several tests by simulation have been performed with the control loop of fig. 59. We shall discuss here just some results achieved with the pH neutralization process.

Using a conventional P-controller with the gain K_R = 1 one gets for the usually in this book as a first test considered case of a jump from pH = 3 to pH = 9 the result shown in fig. 61. The process output is oscillating around pH = 9.

Fig. 62 shows, that already the first run with the Miller control loop gives after some steps an impressive performance improvement and that with 14 repetitions a very good performance is achieved. Not shown results indicate, that the tuning parameter ß should not be selected to be higher than 0,5; ß = 0,3 is a good choice. A change of K_R gives for this case no major modifiactions of the results.

179

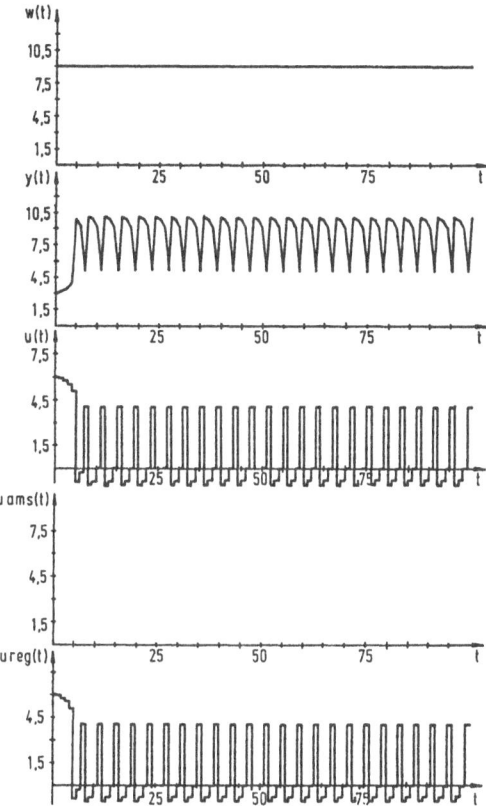

Fig. 61: Basic run with a conventional control loop: a proportional (P-) controller with gain $K_R = 1$, a sampling time $T = 2$ sec and a requested jump from pH = 3 to pH = 9.

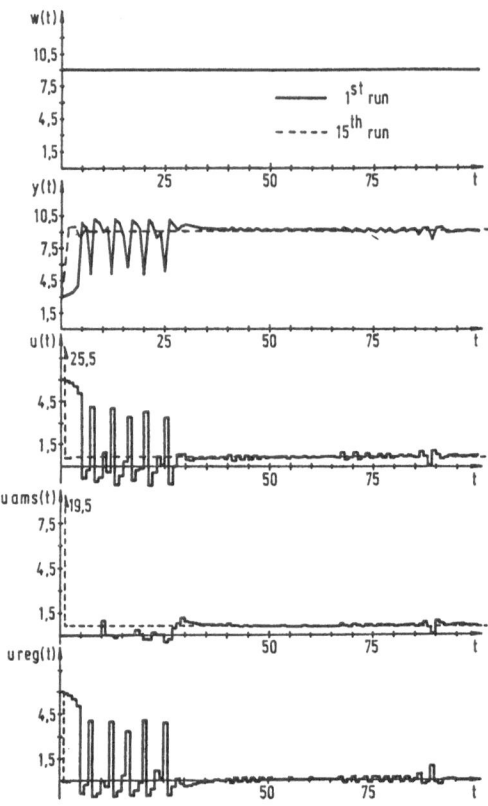

<u>Fig. 62:</u> First run and 15th run with the Miller control loop and a conventional P-controller. All data as in fig. 61, $r^* = 10$ for AMS.

The Miller control loop scheme is also able to handle the cases pH = 8 and pH = 7, however with the necessity to reduce the conventional controller gain drastically (to K_R = 0,1 for pH = 8, to K_R = 0,02 for pH = 7) and with the requirement of much more training (200 runs for pH = 8, 1200 runs for pH = 7) to achieve good results like shown in fig. 62 for pH = 9.

There may arise difficulties if the conventional controller is reacting fairly heavily on not already adapted reactions for the AMS learning device. So it proved to be impossible to achieve a satisfactory performance even for the relatively uncritical case of pH = 9 with a PI-conventional controller, although this controller is itself much better suited for the pH-neutralization process than a P controller. Fig. 63 shows the control loop performance without the u_{AMS} (conventional control loop) and for the 50th run with u_{AMS}: The transient is improved but oscillations are induced which are not present without the learning improvement level.

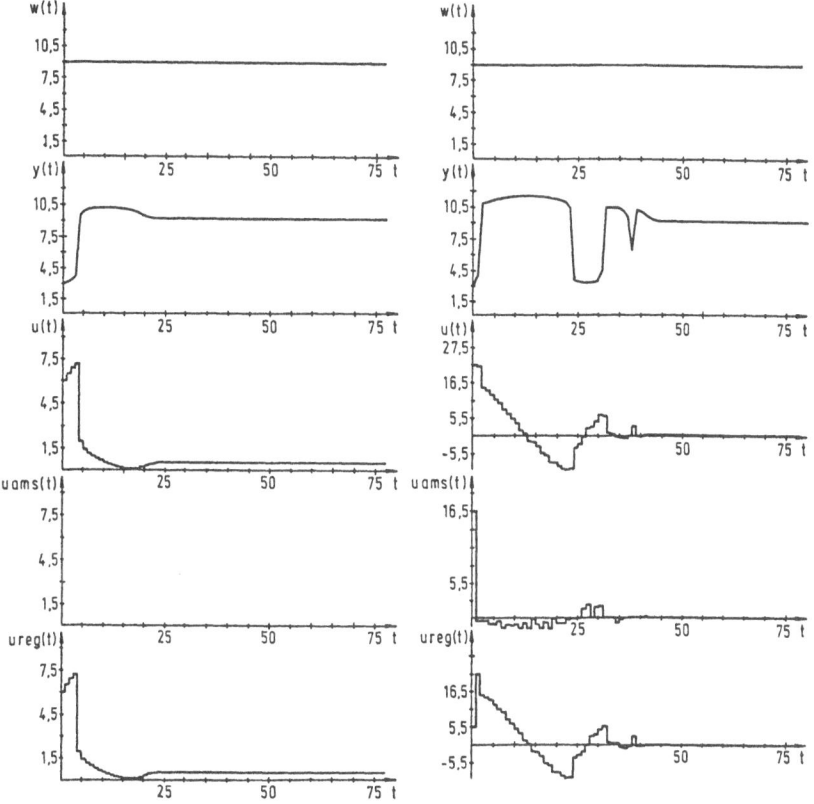

Fig. 63: ph control for pH = 9 with a conventional control loop and a PI controller within -a)- and with the Miller control loop and a PI controller within -b)-. The conventional loop proves here to be the better solution.

However, the Miller learning control loop has on the other hand again the advantage, that the conventional controller implemented helps with unmeasurable disturbances: Fig. 64 shows e.g., that an unmeasured change of the input flow of +/- 50 % can be handled easily by the Miller control loop. The same fact is true for 100 % concentration changes, that means one gets similar advantages as with the Ersü/Wienand approach of fig. 48 - c.f. fig. 53 -.

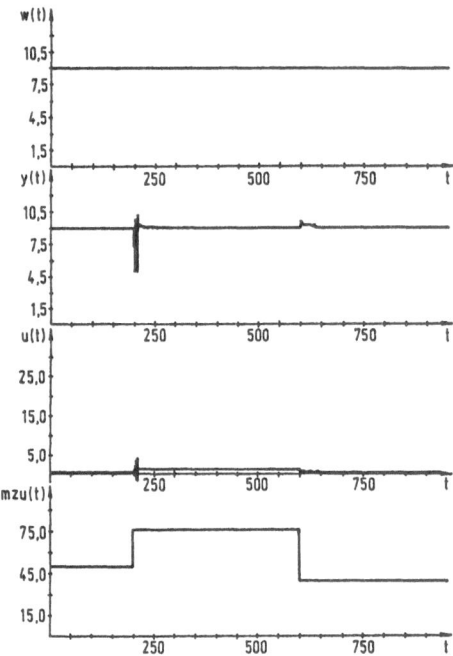

Fig. 64: Reaction of trained Miller control loop to an unmeasured change of the input flow concentration $\dot{m}_i$ by +/- 50 %

In conclusion one can state, that the Miller learning control loop is a little bit more complex to design than LERNAS, since one has to find at first a globally stabilizing controller. Furtheron one cannot make use of multi-step predictions as it is the case in the approach of Ersü/Wienand.

However, the Miller learning control loop is much simpler as the Ersü/Wienand approach and shows the same advantages in unmeasurable disturbance rejection, but needs only one AMS. [2] Therefore it seems advisable to consider the Miller learning control loop always as an interesting control loop candidate, if disturbance rejection is the predominant problem. However, for systems, where multi-step look ahead abilities are important in connection with overall performance optimization, either LERNAS or its Ersü/Wienand extension seem to be the best choice.

---

[2]Actually, M. H. Raibert used already in the early eighties a structure with just one AMS of CMAC-type to learn balancing for a one-leg hopping machine.

III.3. Real-time application of learning control loops to pilot plants

LERNAS and the Miller learning control loop have not only been applied to simulated technical processes but also to some laboratory scale real processes, subsequently called "pilot plants". Such real-time applications make certain rearrangements necessary in the sequence of internal operations. These will be discussed for LERNAS as example first before the results from the tests are presented.

III.3.1. Modifications for real-time, on-line control

For a simulated process one starts - see e.g. (9) and fig. 3 - from some actual situation $[\underline{y}_d(k), \underline{\psi}(k), \underline{v}(k)]$ and computes by optimization $\hat{\underline{u}}^*(k)$ and by active learning the applied process input $\underline{u}(k)$. The computation time is not taken into account, which means that dynamical changes of the process during the time needed for computation are not considered, that is the process is assumed to be practically halted during this time span. For real plants one has now to include into the control loop scheme the fact that the process is dynamically active during all the computations. Actually the following three points have to be handled (/28/):

i.   sampling and process input updating have to occur in a strictly periodic fashion, since the information stored in the predictive process model and the controller is assumed to be arranged in this way. Any change in the time relationships would render the stored information useless.

ii.   considerable delays between measurement data sampling and process input updating instants make the predictive process model very complex because the process situation description by $\underline{y}(k), \underline{y}(k-1) \ldots \underline{y}(k-\alpha); \underline{u}(k), \underline{u}(k-1) \ldots \underline{u}(k-\beta)$ has to go up at least to a $\beta T \geq T_t$ (T being the sampling, T_t being the delay time).

iii. the time delay between sampling of measurement data and its effect on the new process input signal has to be added to the eventual process delay to get the overall delay, the magnitude of which is responsible for system performance degradation in relation to the closed loop performance without delays. Therefore it has to be kept as short as possible.

Regarding the control algorithm as described up to now with respect to points ii and iii it is quickly discovered that its sequencing is not well-suited for a real-time implementation. Examination of the computational effort required for the different tasks within a single computation period shows that most of the time is spent on optimization. Depending on process complexity optimization (including controller updating) consumes 67 % to 99 % of the computation time, active learning 20 % to 0.6 % and the remaining tasks 13 % to 0.4 %. Actually, optimization and active learning need variable time spans, the "remaining tasks" comprising all organisational operations, like hash-coding and so on, however need only a limited time, which is calculable in advance.

Accordingly three measures have been taken:

i. strict time frames are used with ample time calculated in advance, allowed for the remaining tasks. Active learning and/or optimization are adapted to these time frames by stopping the respective calculations when the end of the available time span is reached. Since both operations are already leading to improvements in each of their substeps, this is quite acceptable.

ii. additional data sampling is introduced so as to have at the main sampling instant $k \cdot T$ an appropriate process input $u(k)$ immediately available. The idea behind this is as follows: the sampling interval T has to be long enough to perform at least some optimization steps within it. From this requirement the computer and optimization procedure implementation actually used determines how fast the process to be controlled can be, since during one sampling period no major dynamic process changes are permitted. However, due to the computing time relationships mentioned above, the recall of an appropriate $\hat{u}^{*}$ value from the controller memory plus some active learning needs only a fraction of the optimization time. So at some fraction $\delta_L \cdot T$, $\delta_L \ll 1$, before kT, that means at $(k - \delta_L)T$ a presampling is performed, giving some estimates of $y(k)$, $v(k)$, namely $\tilde{y}(k) = y(k - \delta_L)$, $\tilde{v}_{(k)} = v(k - \delta_L)$.

These values are then used to obtain from (2) some $\tilde{x}(k)$ to find from (9) a suitable $\hat{\tilde{u}}^{*}(k)$. Now active learning can be startet seeking unknown territory in the predictive process model (4) by computing the training status indication η with the help of $\tilde{y}(k)$, $\tilde{v}(k)$, substituting the unknown values $y(k)$, $v(k)$. Since an advantageous first direction, usually given when active learning is applied together with optimization (c.f. section III.1.3), is not available here, a fixed search procedure through all the possible directions is used. The probing step length is always b_o and as an actual $u(k)$ substitution for $\hat{\tilde{u}}^{*}(k)$ the u-value with the lowest η-value is kept in mind. The search is stopped either if one obtains for some direction $\eta = 0$, or if all directions have been tested or if the main sampling instant kT has been reached.

iii. The optimization is postponed until at kT the $u(k)$ found in ii has been applied to the plant. With the measured data at kT the optimization continues until the required accuracy and/or the time instant $[k+1-(\delta_L+\delta_0)] \cdot T$ has been reached. The time period $\delta_0 T$ is required for storage of the found optimal u^{*} and - optionally - data recording, a documentation tool for later examination and behaviour assessment.

The scheduling just described is depicted in more detail in fig. 65. It should be pointed out that the preliminary sampling instant determined by δ_L may be varied during on-line process control, since the computations connected with it are not used for building up the predictive model and/or the controller.

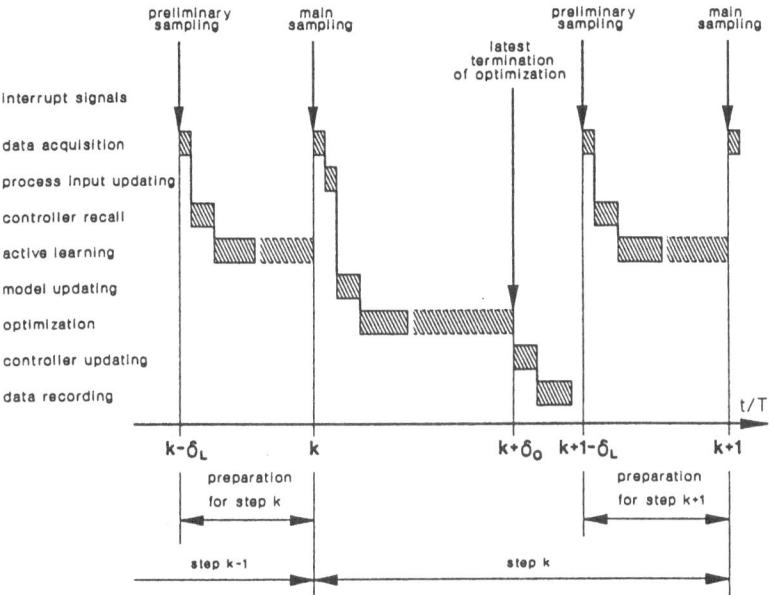

<u>Fig. 65 from /28/:</u> Sampling and computation sequence for real-time, on-line control.
This allows one to improve the quality of the estimates $\tilde{y}(k)$, $\tilde{v}(k)$ by switching off the active learning after some initial learning period, leading to a considerable reduction of δ_L.

By placement of the optimization after process input updating one avoids the respective control action delay. However, we increase the time between new process data acquisition and utilization of this new information for process control, since the computed optimal u^* is applied only if the situation for which it has been created comes up again. This leads to some performance deterioration in the early stages of process control but this levels out if enough process knowledge is built up by learning.

III.3.2. Control of a binary distillation column pilot plant with LERNAS

In chemical process engineering very often a three step development cycle is followed: The first step
is a laboratory scale process investigation to study key questions. The second step is than a medium
size simulation of major elements of the real plant, this step being used for gaining knowledge in
scaling up the laboratory size process and for pre-optimization of this technical lay-out. Only after
these two steps have been successfully performed is the real plant built up in a size adequate for the
envisaged amount of production. Normally the medium size process, called in the following the
"pilot plant", is kept in parallel with the process to analyse reasons for unexpected behaviour and
to test new ideas of performance improvement without the necessity of interrupting the production,
as would be the case if these studies were performed on the real plant.

The binary distillation column pilot plant used for testing real-time, on-line application of
LERNAS was such a pre-investigation, research and process optimization facility of laboratory size
owned by the German company Bayer AG [3]. The main element was a column of 1300 mm length
filled with solid particles imitating a production plant with 13 trays.

The control objective is to regulate the bottom and top compositions simultaneously, which means
a multivariable control task has to be undertaken. Due to the fact that the pressure (approx.
990 mBar) is held constant by an independent external control loop, the compositions can be
monitored indirectly by the temperatures y_1 being the value at the bottom and y_2 being the value
at the top. Control is applied to the column by the heating (i.e. heating power) at the bottom u_1
and the reflux flow-rate u_2 at the top. The feed flow-rate of the two-product composition (45 %
and 55 % respectively) was 2 l/h (liter per hour) at a feed-temperature of approximately $1000°$ C.
Table 4 gives the considered operating space and the chosen quantization for each variable.

	Min	Max	ϵ
y_1	$100°$ C	$150°$ C	$0,05°$ C
y_2	$100°$ C	$150°$ C	$0,05°$ C
u_1	0 W	640 W	2 W
u_2	0	4	0,025

Table 4 - Ranges and quantization for plant outputs and inputs, example distillation column.

---

[3]The application was made possible by the active interest taken in our work by Professor
Dr. M. Polke and Dr. J. König from the BAYER AG.

Studies with different performance criteria and LERNAS parameter settings were performed, from which results of the following particular case will be presented:

- $T = 35$ sec (chosen from process step response observations)

- $$J(k) = (w_1 \cdot y_1, w_2 \cdot y_2)|_k \cdot \begin{bmatrix} \dfrac{10^4}{y_{1max} \cdot y_{1min}} & 0 \\ 0 & \dfrac{10^4}{y_{2max} \cdot y_{2min}} \end{bmatrix} \cdot \begin{bmatrix} w_1 \cdot y_1 \\ w_2 \cdot y_2 \end{bmatrix}\Big|_k +$$

$$+ \left[u_1(k) \cdot u_1(k-1), u_2(k) \cdot u_2(k-1) \right] \cdot \begin{bmatrix} \dfrac{15}{u_{1max} \cdot u_{1min}} & 0 \\ 0 & \dfrac{4}{u_{2max} \cdot u_{2min}} \end{bmatrix} \cdot \begin{bmatrix} u_1(k) \cdot u_2(k-1) \\ u_2(k) \cdot u_2(k-1) \end{bmatrix}$$

- M_P: $[\underline{y}(k), \underline{y}(k-1), \underline{u}(k)] \longrightarrow \underline{y}(k+1)$; $r_P^* = 16$

  M_C: $[\underline{w}(k) \cdot \underline{y}(k), \underline{y}(k), \underline{y}(k-1)] \longrightarrow \underline{u}(k)$; $r_C^* = 8$

Since the pilot plant, being an active element in the day-to-day work of the chemical company, was available only for a limited time span, and since chemical plants are relatively slow plants, some strategy to minimize learning at the plant was appropriate. Two procedures are possible: either to collect data from the plant and use it for building up a rough predictive model, with which a preliminary control strategy can be developed in an accelerated manner or to bring into the controller associative memory some preliminary information by controlling the process manually and feeding the controller memory in parallel with this information. The latter approach was used in connection with the distillation column and is the time period before the letter H in the diagrams of fig. 66.

After this "H" LERNAS has taken over control of the plant. The improvement of the marked settling times demonstrates the efficiency of LERNAS. Similar results were achieved with 10 % and 20 % feed-flow disturbances (rate and composition), where after a brief learning period the control eliminates the disturbance effects on the steady state values successfully.

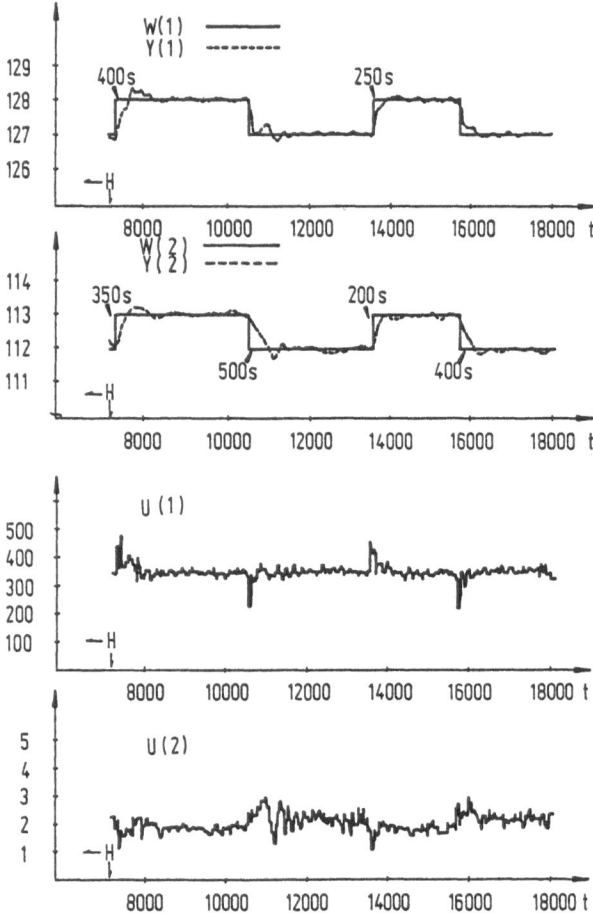

Fig. 66: Control of a distillation column pilot plant. Till the letter H manual control was used. From thereon LERNAS took over. One sees the improvement in the time span to reach the desired new values w_1, w_2 after a set point change. 400 sec → 250 sec, 350 sec → 200 sec.

Furthermore, some comparisons were made with an adaptive control concept implemented already on the distillation column pilot plant. The method used was Koivo's (/29/) multivariable generalization of a self-tuning single-input, single-output controller from Clarke and Gawthrop (/30/). Since LERNAS and the adaptive control could both handle the distillation column in a satisfactory way, no essential control characteristic differences were to be expected. The results of the comparisons are in accordance with this expectation. In both cases the deviations from the desired value after a set point change quickly became negligible. The self-tuning controller showed some advantages in leading to a smoother behaviour of the inputs, LERNAS in being able to suppress overshoot fully, since with the self-tuning controller a final amount of 0,6% overshoot remained. However, these differences are so small, that they could be certainly levelled out by tuning the free parameters of the adaptive and/or the learning scheme in an appropriate way.

Actually, clear distinctions in performance between adaptive and learning control can be anticipated only if at least one of these control schemes operates at the boundary of its abilities. The
distillation column control seems to be an easy task for both schemes, so the question of a comparison will not be pursued here further.

III.3.3. Control of an air conditioning system with Miller's approach

Actually, it would have been interesting to apply the Miller learning control loop concept to the
distillation column also. However, when we started to look at simplifications and extensions of
LERNAS, we no longer had access to this plant. But there are a number of other processes in real
hardware available at the Institute of Control Engineering of the TH Darmstadt, which are run for
educational and research purposes by the Control Systems and Process Automation Section (Prof.
Dr.-Ing. e.h., Dr.-Ing. R. Isermann).

One of these processes is an air-conditioning plant. It has been used especially for testing adaptive
control concepts, and I am indebted to my colleague Professor Isermann, for making the plant
available to us, although it was needed in parallel for research work of his own group.

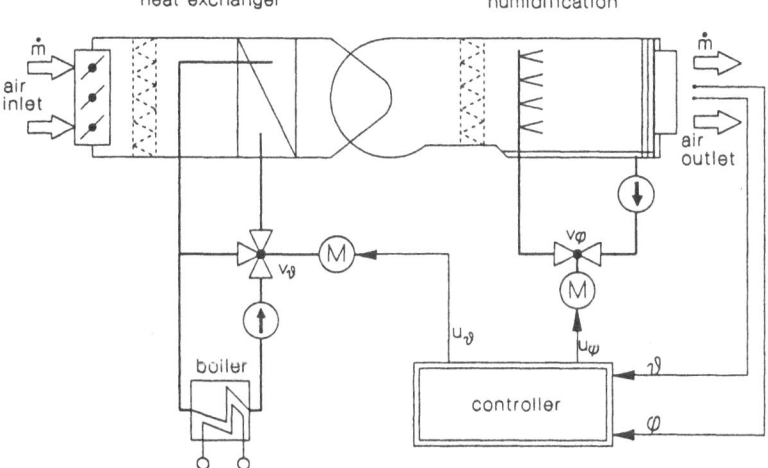

Fig. 67 (from /31/): Schematic sketch of the pilot plant "air conditioner"; $\dot{m}$ air flow, ϑ temperature, φ relative humidity, V_ϑ / V_φ valves, M electric motors, u_ϑ, u_φ voltages = motor inputs.

The air-conditioner allows one to control simultaneously the air-temperature and humidity, and it is also fairly non-linear. Fig. 67 shows a schematic sketch of the plant, and at first a short description will be given, based on /31/: With the help of a blower a certain volume of air is driven through the system, the throughput being selectable between $\overset{\bullet}{m} = 0\ m^3/h$ and $\overset{\bullet}{m} = 500\ m^3/h$. The air is warmed by hot water in the heat exchanger. The amount of hot water used can be regulated by a split-value V_ϑ which is manipulated through the input voltage u_ϑ to an electric motor M. With rising temperature one has a decrease of the relative air humidity, since the absolute air humidity stays constant (compare the Mollier-diagram of fig. 68, line AB).

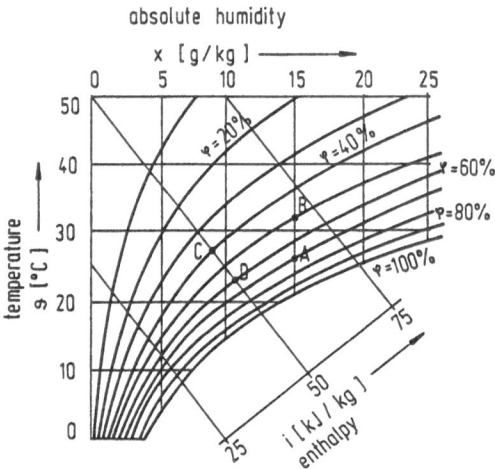

Fig. 68 (From /31/): Mollier-diagram for the air-conditioner; x = absolute humidity, φ = relative humidity, ϑ = temperature, i = entalphy.

To influence the air-humidity water is injected into the air-stream by six spray-nozzles. The amount of water injected is regulated by the value V_φ, driven by an electric motor with the control input u_φ. With u_φ the absolute and the relative humidity increase and the air temperature decreases, e.g. along the line of constant entalphy CD in the Mollier-diagram - fig. 68 -, if the humidification takes place just at the cooling border, which means that no energy is fed into the air-stream. At the air outlet two sensors are installed, one for measuring the temperature and one for measuring the relative humidity, so that the control loop can be closed with an appropriate two-input, two-output controller.

If one describes the dynamic behaviour of the two-input $(u_\vartheta,\ u_\varphi)$ - two output $(\vartheta,\ \varphi)$ process by a P-canonic process model, one obtains from the behaviour described above, that for positive gains $k_{\vartheta\vartheta}$ and $k_{\varphi\varphi}$ in the channels $u_\vartheta -> \vartheta$, $u_\varphi -> \varphi$ there are negative gains $k_{\vartheta\varphi}$ and $k_{\varphi\vartheta}$ in the couplings, as indicated in fig. 69.

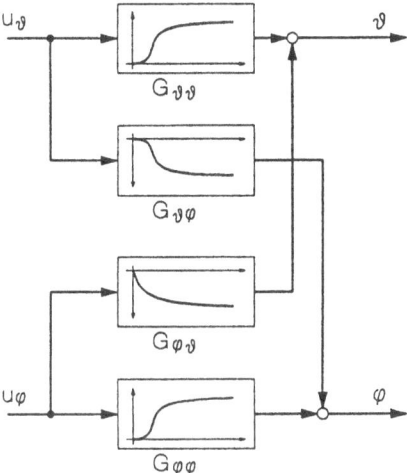

Fig. 69 (from /31/): P- canonic block- diagram for the air- conditioner.

The overall static and dynamic behaviour is non- linear, due to the non- linear characteristics of the valves, the heat exchange process in the heater and the water spray process for humidification. It is furthermore highly dependent on the air stream, $\overset{\bullet}{m}$, which has to be considered here as a distur- bance (the main one in fact). An increase of $\overset{\bullet}{m}$ decreases the temperature, since more air has to be warmed up in a given time. The relative humidity, on the other hand, can go up or down, depend- ing on the correct working situation: e.g. if the humidity of the air is already correct, so that no water at all has to be injected, an increase of the air flow, $\overset{\bullet}{m}$, means an increase of the relative humidity also, since the temperature goes down but the absolute humidity stays constant (compare fig. 68). However, if water injection is necessary to obtain a certain absolute humidity level, an increase of $\overset{\bullet}{m}$ implies that more air has to be humidified in a given time. That means, the absolute humidity goes down in parallel with the temperature, which may also lead to a decrease of the relative humidity (compare again fig. 68).

For controlling the air conditioning system by the Miller learning control concept the following selections were made by M. Hormel and K. Kleinmann, who performed this work:

1.) Sampling time: 45 sec

2.) Conventional Controller: PI- Controllers for ϑ, φ without coupling elements:

$$u_{\vartheta}(k) = u_{\vartheta}(k-1) + 4{,}3 \, [w_{\vartheta} - \vartheta]_{|k} - 3{,}87 \, [w_{\vartheta} - \vartheta]_{|k-1}$$

$$u_{\varphi}(k) = u_{\varphi}(k-1) + 2{,}25 \, [w_{\varphi} - \varphi]_{|k} - 1{,}575 \, [w_{\varphi} - \varphi]_{|k-1}$$

3.) Learning control level: AMS

AMS- characteristics: $r^* = 16$

response construction as in fig. II.12.b

weight change procedure as given by equat. II.7.

number of inputs 10

number of outputs 2.

Input- stimuli	Scaling b (fig. II.10)	Resolution ϵ
$w_\vartheta(k)$	32	0,03° C
$\vartheta(k)$	20	0,05° C
$\vartheta(k-1)$	"	"
$u_\vartheta(k-1)$	10	0,1 V
$u_\vartheta(k-2)$	"	"
$w_\varphi(k)$	16	0,06 %
$\varphi(k)$	10	0,1 %
$\varphi(k-1)$	"	"
$u_\varphi(k-1)$	10	0,1 V
$u_\varphi(k-2)$	"	"

4.) Parameters for the updating of the additional controls $u_{AMS\vartheta}, u_{AMS\varphi}$ (in equat. (61)):

$\beta_\vartheta = 0,99$; $K_{R\vartheta} = 4$; $\beta_\varphi = 0,99$; $K_{R\varphi} = 2.$

5.) Air throughput $\dot{m} \approx 200 \text{ m}^3/\text{h}$.

The basis of these selections has been the knowledge about the general process behaviour and advance trials to obtain numerical values for the parameters. However, the results are fairly robust with respect to these selections.

For the experiments with the pilot plant profiles of the desired temperature and relative humidity values w_ϑ and w_φ had to be chosen. This was done by taking into account the limitations of reachable ϑ, φ- Values due to the following bounds on the valves and/or valve motor input voltages:

V_ϑ ; (u_ϑ) min 10 % of the overall valve- lift (input voltage)

                  max 80 % of the overall valve- lift (input voltage)

V_φ ; (u_φ) min 25 % of the overall valve- lift (input voltage)

                  max 75 % of the overall valve- lift (input voltage).

Fig. 70 shows the field of ϑ, φ - values allowed by these bounds on the valve-lifts and the w_ϑ, w_φ - values selected on this basis. The trace should not be misinterpreted as a time profile. Between the marked points either jumps or changes in a ramp form were performed. Details can be found in the figures 71, 72.

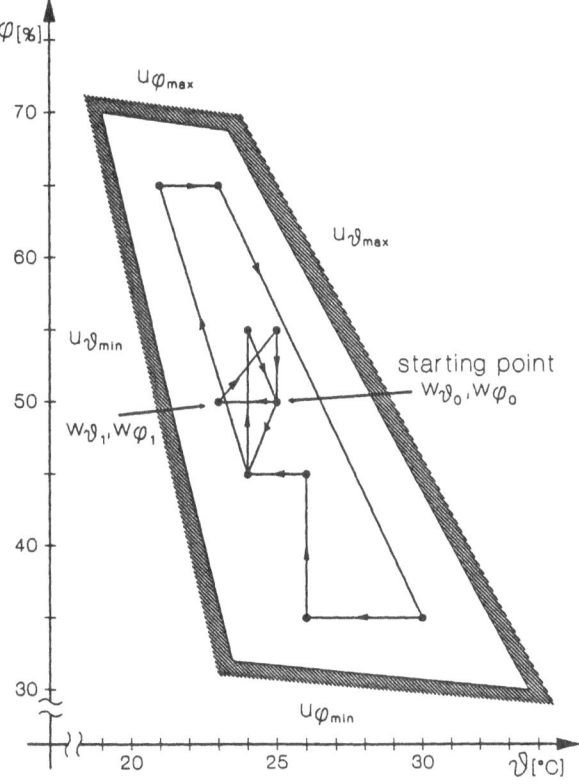

Fig. 70: Limitations on reachable ϑ, φ-values due to bounds on the valve-lifts of the used air-conditioning system and trace of the accordingly selected w_ϑ, w_φ-values.

To provide a reference for the possible improvements using a learning control, the result of controlling the air conditioning system by the two decoupled PI-controllers only is shown in fig. 45. One can see that the task is solvable for the PI-controllers (the basic assumption with respect to the conventional controllers in the Miller concept), but that, especially for the temperature, the control is not fully satisfactory. Fig. 72 gives the results of the first run with the learning level being active. Surprisingly, already in this first run an improvement with respect to control behaviour is achieved, as can be seen e.g. in the plot of the temperature after the temperature jump at t = 1 h, the temperature jump at t = 5,5 h and the relative humidity jump at t = 6,25 h. To explain this, one has to take into account, that the time axis in the figures 71, 72 is very much compressed in relation to the sampling time of 45 sec.

194

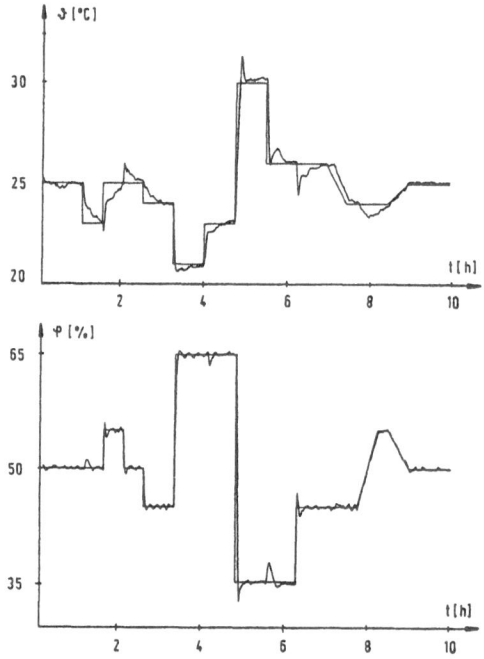

Fig. 71: Air conditioner control by two decoupled PI-controllers; a-temperature, b-relative humidity.

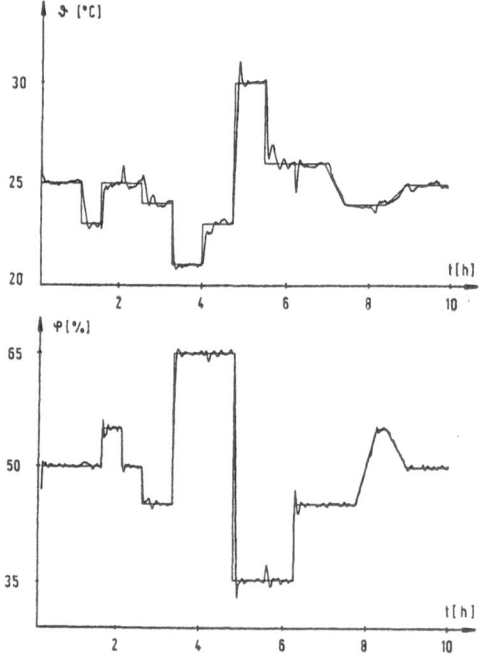

Fig. 72: Air conditioner control by the Miller control loop, first run with the learning level active; a-temperature, b-relative humidity.

During the decay of the temperature after the temperature jump at 1 h e.g. 25 - 50 sampling steps and/or learning steps occur. Fig. 73 supports this remark by showing the voltage input u_ϑ for the split-valve, which regulates the heat exchanger behaviour, for the control cases without and with the learning level (fig. 71, 72). One sees that especially for the two situations mentioned above the input is heavily influenced by additional signals from the upper level, which are added to the signals of the PI controller.

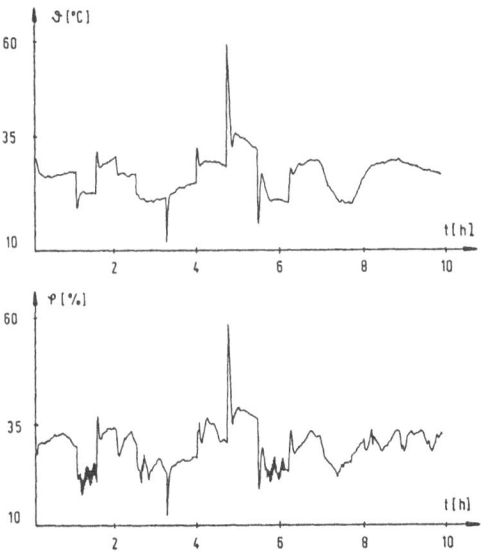

Fig. 73: Voltage input u_ϑ for the split-value which regulates the heat exchanger behaviour; a - for the case without learning (cf fig. 71), b - for the first run with the learning level active (cf. fig. 46).

Fig. 74 shows, that repetitions of running the plant with the demanded temperature and relative humidity profiles bring some further improvements in the control behaviour. However, one reaches quickly a limit for further improvements in view of the plant inertia, the measurement noise and the inevitable inaccuracy of the actuators.

In addition to the discussion of the improvements over conventional control some comparison with adaptive control as an alternative approach to handle fairly unknown, non-linear processes is of interest. Unfortunately we were unable to use the results and the software from /31/ directly, since some changes in the plant hardware and the process controlling computer hardware took place in the period between R. Schumann's research work and the experiments by our own group. However, the identification method/controller approach selection as well as the software program used by us are based on advice given by R. Schumann to J. Militzer, who looked into the adaptive control problem, so that we feel that the adaptive control tools used are adequate for the task considered. We are very much indebted to R. Schumann for his advice.

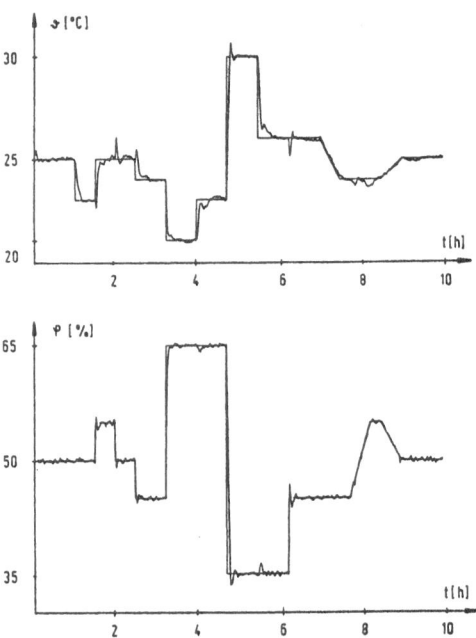

Fig. 74: Air conditioner control by the Miller control loop, fifth run with the learning level active; a- temperature, b- relative humidity.

One of the results of /31/ is, that for multi- input, multi- output adaptive control a good approach, in view of performance and computation time, is to apply a P- canonic input/output model, of which the parameters are estimated by the recursive least squares method, together with a matrix- polynomial dead- beat controller. For details see /31/. In connection with the air- conditioning system R. Schumann suggested that J. Militzer should work with this combination by using submodels of the order three and a forgetting factor for adaptation of 0,95. Furtheron, to reduce the amplitudes of the actuator movements, for the dead beat controller one control step more than the plant model order was implemented.

Fig. 75 shows the results of adaptive control for the air conditioning system generated on this basis, with the same temperature and relative humidity requirements profile as used already for fixed PI- control and for control with the Miller learning control loop. After a certain period necessary for model parameter estimation the control task is performed fairly well. However, looking at the more critical temperature control one sees that practically nothing is gained in comparison with the fixed PI- controllers, (the plant seems not to be a problem for fixed PI- controllers contrary to the expectations gained from /31/), and that the learning control loop performs better. The problems with the ramp at t = 8,5 h are due to the fact that the adaptive controller requires here higher actuation than the actuator limitations allow - (see fig. 76).

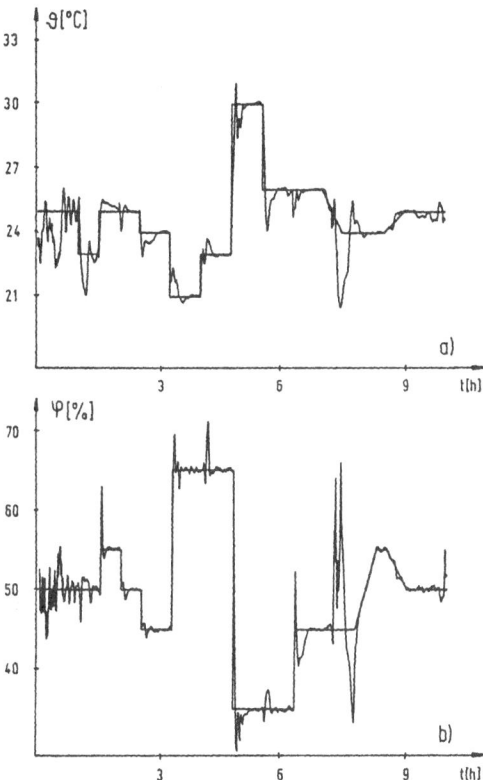

Fig. 75: Air conditioner control by an adaptive approach based on /31/; a- temperature, b- relative humidity - same requested temperature relative humidity course as used in fig. 71, 72 and 74.

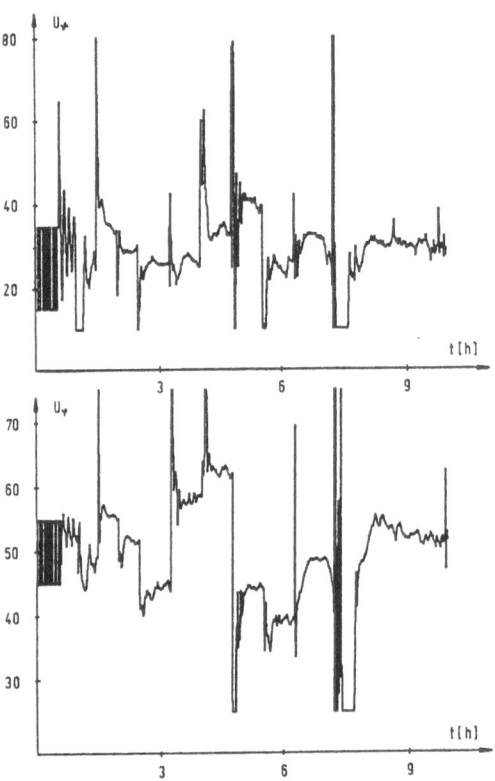

<u>Fig. 76:</u> Voltage inputs u$_\vartheta$, u$_\varphi$ for adaptive control and the demanded temperature, relative humidity profiles of fig. 75. The actuators run partly against their limits.

It should be remarked that the results shown for all the control approaches considered here are typical examples from a number of experiments run using the air conditioning system.

Therefore one can state, as a conclusion that the Miller learning control loop is able to perform real-time control of real multi-input/multi-output plants as well as Ersü's learning control loop LERNAS.[4] It is able to improve the performance of adequate fixed linear controllers and, for the air conditioner, outperforms adaptive control, at least for the adaptive control approach used.

III.4. Recapitulation

In this third main chapter the learning control loops considered by my group have been expounded together with the performances which can be achieved by these structures. After a thorough discussion of the basic learning control loop layout "LERNAS" the test processes have been put forward, which have been used mainly for performance evaluation. The results found by simulation with respect to the ability to control processes only crudely known have been shown, as well as results of investigations regarding sensitivity with respect to control loop parameters to be selected in advance of LERNAS application and of investigations on closed loop stability, control of integrating, dead-time comprising and non-minimum behaviour displaying plants. Then we have looked at hierarchies with and alternatives to LERNAS, demonstrating, that on one hand a vast variety of learning structures can be postulated and are working in a satisfactory way, too, but that on the other hand for such macro-structures also some restrictions exist due to their detailed specification. Not discussed was the point that knowledge about the process and/or a suitable basic controller can be implemented instead by prestructuring the interpolating memories by just implementing adequate elements in parallel to the interpolating memories, so that these devices learn only improvements to already known basic approaches. Finally for LERNAS and the most simple on-line learning control structure, the Miller learning control loop, the on-line use for the control of real pilot plants has been discussed and it has been shown by the respective results, that the learning control loops are sufficiently developed to be realistic tools for the control engineer.

III.5. Literature:

/1/   Militzer, J.; Tolle H.: Prüfung der Eignung von assoziationsfähigen Modellen der menschlichen nervösen Informationsverarbeitung zur selbsteinstellenden Regelung eines technischen Prozesses. Zwischenbericht zum DFG-Forschungsvorhaben To 75/9-1, Dez. 1982

/2/   Tolle, H.: Optimization Techniques, Springer Verlag, 1975

/3/   Pierre, D.A.: Optimization Theory with Applications, John Wiley, 1969

---

[4]The relative advantages and disadvantages of the Miller concept, LERNAS and other concepts have been discussed already at the end of section III.2.5.

/4/   Hooke, R.; Jeeves, T. A.: Direct Search Solution of Numerical and Statistical Problems, Journ. of the Assoc. for Comp. Machinery 8, 1981

/5/   Schwefel, H.-P.: Numerische Optimierung von Computermodellen mittels der Evolutionsstrategie Birkhäuser Verlag, 1977

/6/   Ersü, E.; Mao, X.: Control of pH by use of a self-organizing concept with associative memories. Preprints Int. IASTED Conference on "Applied Control and Identification" Copenhagen, 1983

/7/   Asselmeyer, B.: Zur optimalen Endwertregelung nicht-linearer Strecken mit Hilfe kleiner Rechner Dissertation, T.H. Darmstadt, 1980

/8/   Douglas, J.M.: Process Dynamics and Control, Volume 1 - Analysis of Dynamic Systems. Prentice - Hall Inc., 1972

/9/   Rosenbrock, H.H.; Storey, C.: Computational techniques for chemical engineers. Pergamon Press, 1966

/10/   Ruppert, M.: Reglersynthese mit Hilfe der mehrgliedrigen Evolutionsstrategie. VDI-Verlag - Fortschrittsberichte VDI-Z, Reihe 8, Nr. 81, Düsseldorf, 1982

/11/   Militzer, J.: Prüfung der Eignung von assoziationsfähigen Modellen der menschlichen nervösen Informationsverarbeitung zur selbsteinstellenden Regelung eines technischen Prozesses. 2. Zwischenbericht 1.1.83 - 31.12.83, DFG-Vorhaben To 75/9-3, T.H. Darmstadt, 1983

/12/   Saridis, G.N.: Self-Organizing Control of Stochastic Systems. Control and Systems Theory, Vol. 4 Marcel Dekker Inc. New York, 1977

/13/   Militzer, J.; Tolle H.: Vertiefungen zu einem Teilbereiche der menschlichen Intelligenz imitierenden Regelungsansatz. Proceedings-Jahrestagung der deutschen Gesellschaft für Luft- und Raumfahrt München, 1986

/14/   Agathoklis, P.; Burton, L.T.: Practical BIBO-Stability of n-dimensional discrete systems. IEEE Int. Symp. on Circuits and Systems, Newport Beach, USA, 1983

/15/   Ersü, E.; Gehlen S.: Vertiefende Untersuchungen zu gemäß der menschlichen nervösen Informationsverarbeitung lernenden Systemen. Abschlußbericht für das von der VW-Stiftung geförderte Vorhaben I/60/901, T.H. Darmstadt 1988

/16/   Dörner, D.: Problemlösen als Informationsverarbeitung. Verlag H. Huber, 1974

/17/   Ersü, E.; Tolle H.: Hierarchical learning control - An approach with neuron-like associative memories. Collected papers of the IEEE Conf. on Neural Information Processing Systems, Denver, USA 1987 - Published by the American Institute of Physics, 1988

/18/   Paul, R.P.C.:Robot Manipulators: Mathematics, Programming and Control. MIT Press Cambridge, MA, 1981

/19/   Raibert, M.H.: Analytical equations versus table look-up for manipulation: A unifying concept. IEEE 8th Conference on Decision and Control, New Orleans, Dez. 1977

/20/   Davison, D.J.; Goldenberg, A.: Robust control of a general servomechanism problem: The servo compensator. Automatica, 1975

/21/   Tolle, H.: Mehrgrößenregelkreissynthese Bd. II: Entwurf im Zustandsraum, Oldenbourg Verlag, 1985

/22/   Ersü, E.; Wienand S.: An associative memory based learning control scheme with PI-controller for SISO-nonlinear processes. IFAC Symposium on Microcomputer Applications in Process Control Istanbul (Turkey), 1986

/23/ Tolle, H.; Ersü, E.: Proposal for the research grant regarding "Vertiefende Untersuchungen zu gemäß der menschlichen nervösen Informationsverarbeitung lernenden Systemen -/15/-", Darmstadt 1985

/24/ Miller, W.T. III; Hewes. R.P.: Real time experiments in neural network based learning control during high speed nonrepetitive robotic operations. Proc. 3rd IEEE Int. Symp. on Intelligent Control, Washington, DC (USA), 1988

/25/ Miller, W.T. III; Glanz, F.H.; Kraft, L.G. III: Application of a general learning algorithm to the control of robotic manipulators. The International Journal of Robotics Research, Vol. 6, No. 2, 1987

/26/ Ersü, E.; Tolle, H.: Acceleration of learning by variable generalization for on-line, self-organizing control.Vth Polish-English Seminar "Real-Time Process Control" Radziejowice (Poland), 1986

/27/ Ersü, E.; Tolle H.: A new concept for learning control inspired by brain theory. Proc. 9th IFAC World Congress, Budapest, 1984

/28/ Ersü, E.; Militzer, J.: Real-time implementation of an associative memory-based learning control scheme for non-linear multivariable processes. The Institute of Measurement and Control Symposium "Application of Multivariable System Techniques", Plymouth, England, 1984

/29/ Koivo, H.N.: A Multivariable self-tuning controller, Automatica 16, 1979

/30/ Clarke, D.W.; Gawthrop, P.J.: Self-Tuning Control,Proceed. IEE 126, 1979

/31/ Schumann, R.: Digitale parameteradaptive Mehrgrößenregelung - Ein Beitrag zu Entwurf und Analyse - Dissertation der T.H. Darmstadt, PDV-Berichte, KfK-PDV 217, Kernforschungszentrum Karlsruhe, 1982.

Conclusion and remarks on further research topics

Compared with the international effort of the last two decades in the area of adaptive control, the total work on learning control is very small. Therefore, one cannot expect, that the respective status is equivalent to the achieved status of adaptive control. However, a renewed interest in imitation of human abilities has arisen with J.J. Hopfield's establishment of a close analogy between the spin glass theory from physics and neural network modelling (/1/). Although the main impact of the respective studies goes in the direction of parallel computation and adequate computer architectures, some stimulation of research on learning systems has taken place, too. It was stated already in the preface, that I believe that a simulation on neuronal level may be a more detailed modelling than actually needed for imitating the flexible behaviour of biological systems. The main point is in my opinion that by locally generalizing memorization units general nonlinear mappings can be learned, something which seems to be difficult to achieve with usual adaptive methods, by the way. As an example for this claim, that just the automatically interpolating non-linear mapping is of importance, one may use the take over of human operator strategies displayed in fig. II.29, where human behaviour is learned without using one of the standard neural network models. In fact, the more condensed modelling of information storage used in this book requires less training and/or learning steps for acquiring knowledge than neural network models need, in general.

It has been pointed out in the book in several places, that a number of questions are still open. Up to now, the amount of history characterizing the current situation has to be fixed in advance. Is this necessary, or can one think of a procedure, by which an optimal choice is made automatically, like the process order selection in adaptive control? With respect to dead-times and non-minimum phase behaviour, is this something, which human beings learn by trial and error without any special mental effort, or is for processes with such properties a conscious recognition of an unusual behaviour necessary, which has to be modelled by some context input and/or structural change for the learning control loop? Also the studies on learning control loop hierarchies are still in their infancy and for the different macro-structures and especially macro-structure simplifications some further research work seems to be appropriate.

However, the biggest challenge lies, in my opinion, in another direction(c.f. /2/). Learning control loops imitate to a certain extent just acquired abilities of senso-motoric coordination. Human beings are in addition able to make trade-offs and decisions on a symbolic level, which is modelled by the so-called artificial intelligence methods. Now, both the symbolic reasoning and the ability level are closely connected. For reasoning on abilities, the acquired abilities have to be recognized as a certain entirety and thus reduced to some object in the symbolic world representation. Furthermore, one knows that abilities are often learned more easily by spoken orders, e.g. of a sports teacher, which means by additional spoken inputs. So modelling the human overall efficiency to deal with its environment requires to integrate learning control and artificial intelligence methods into one overall intelligent system. How this can be done, which and how much structure has to be specified, how far eventually besides of information also structural connections can be modified by learning and what may be an appropriate, complex enough task to verify such a system, seem to me to be the main questions to be addressed in future.

Literature

/1/ Hopfield, J. J.: Neural Networks and Physical Systems with Emergent collective Computational Abilities. Proc. Natl. Acad. Sci. USA, Vol. 79, 1982

/2/ Tolle, H.: Autonomiererhöhung durch Imitation menschlicher Intelligenz, Automatisierungstechnik (at). 1991

Index

Lecture Notes in Control and Information Sciences

Edited by M. Thoma and A. Wyner

Lecture Notes in Control and Information Sciences

Edited by M. Thoma and A. Wyner